The UK Mathematics Trust

Yearbook

1999 – 2000

This book contains an account of UKMT activities from 1st September 1999 to 31st August 2000. It contains all question papers, solutions and results as well as a variety of other information.

Published by the United Kingdom Mathematics Trust.
School of Mathematics, The University of Leeds, Leeds LS2 9JT
Telephone: 0113 233 2339
Website: http://www.mathcomp.leeds.ac.uk/

Cover design: – The backdrop is a Penrose tiling whose complexity reflects the activities of the UKMT.

The photographs are

Front Cover: IMC success at Abraham Darby School, Telford

Back Cover: Girls of Burgess Hill School
and
pupils from Peebles High School
celebrating their UKMT success

ISBN 0 9536823 1 5

Contents

Foreword i

Background and history of the UKMT 1

Foundation of the Trust; Aims and Activities of the Trust; Structure and Membership of the Trust

An outline of the events 3

Junior competitions; Intermediate competitions; Senior competitions

The Junior Mathematical Challenge and Olympiad

- JMC paper and solutions 6
- JMC answers and comments 13
- JMO paper and solutions 17
- JMO marking and results 25

The Intermediate Mathematical Challenge and its follow-up events

- IMC paper and solutions 28
- IMC answers and comments 36
- Kangaroo paper and solutions 39
- Kangaroo awards and high scorers 47
- IIIMC Grade 10 paper and solutions 50
- IIIMC Grade 11 paper and solutions 55
- Comments on IIIMC responses, awards and high scorers 59
- National Mathematics Summer School 64

The Senior Mathematical Challenge and British Mathematical Olympiads

- SMC paper, solutions and high scorers 71
- BMO Round 1 – paper and high scorers 84
- BMO Round 2 – paper and high scorers 90
- BMO Round 1 – solutions 93
- BMO Round 2 – solutions 102
- Olympiad Training Session, Trinity College, Cambridge 107
- 41st IMO, Taejon, South Korea 111

Other aspects of the UKMT and other similar bodies overseas 134

List of volunteers involved in the UKMT's activities 139

Foreword

The UKMT yearbook was published for the first time in the autumn of 1999 and was such a success that it established itself immediately. We welcome you here to the second volume. As usual (by which I mean as last year and as you can expect in the future) you will find here a description of the Trust and a record of its main activities during the school year 1999/2000. For each competition run by the sub-trusts of UKMT we publish the question paper, specimen solutions, and information about the outcomes. You will find here some challenging, interesting, entertaining and educative problems. If you have not seen them before then you are cordially invited to test your mathematical skills and insight by ignoring our answers and finding your own. Good mathematical problems can usually be approached in several different ways and we hope and believe that you will find alternative solutions to those suggested here, some of which should be better than ours.

The UKMT, as a charitable organisation, has just one employee of its own. This is Angela Gould, Executive Director, whom we warmly welcomed in May 2000. The major administrative work of running the Challenges is done on our behalf by the Maths Challenges Office of the University of Leeds. All who have come in contact with Heather Macklin and Jenny Gill will know how effective and friendly they are. The remainder of the work, setting the Challenge and Olympiad papers, marking the Olympiads, training the international Olympiad team, organising all the follow-on competitions: all this is done by volunteers and enthusiasts working in their own time. I feel confident that you would join me in expressing publicly our very warm thanks to all who are involved.

Peter Neumann

Queen's College, Oxford: November 2000.

Background and history of the UKMT

Foundation of the Trust

National mathematics competitions have existed in the UK for several decades. Up until 1987 the total annual participation was something like 8,000. But since then there has been an enormous growth – from 24,000 in 1988 to around a quarter of a million in 1995 – without doubt due to the drive, energy and leadership of Dr Tony Gardiner. By the end of this period there were some nine or ten competitions for United Kingdom schools and their students organised by three different bodies: the British Mathematical Olympiad Committee, the National Committee for Mathematical Contests and the UK Mathematics Foundation. During 1995 discussions took place between interested parties which led to agreement to seek ways of setting up a single body to continue and develop these competitions and related activities. This led to the formation of the United Kingdom Mathematics Trust, which was incorporated as a company limited by guarantee in October 1996 and registered with the Charity Commission. The Royal Institution of Great Britain accepted appointment as its Parent Body. The Mathematical Association became a Participating Body and the Association of Teachers of Mathematics, the Edinburgh Mathematical Society, the Institute of Mathematics and Its Applications, the London Mathematical Society and the Royal Society became Supporting Bodies.

Aims and Activities of the Trust

According to its constitution the Trust has a very wide brief, namely "to advance the education of children and young people in mathematics". To attain this it is empowered to engage in activities ranging from teaching to publishing and lobbying. But its focal point is the organisation of mathematical competitions, from popular mass "challenges" to the selection and training of the British team for the annual International Mathematical Olympiad (IMO).

There are three main challenges, the UK Junior, Intermediate and Senior Mathematical Challenges, which attract over a third of a million entries each year from thousands of schools. These are organised by the Junior and Intermediate Challenges Subtrust (JICS) for pupils aged between 12 and 16 approximately, and the Senior Challenges Subtrust (SCS) for pupils aged 16 or over. They are open to all comers of the appropriate age. Gold, silver and bronze certificates are awarded for the best performances and the most successful participants are encouraged to enter follow-up competitions.

At the junior and intermediate levels the follow-on competitions have entries ranging from around one to three thousand. They are organised by the Junior Olympiad Subtrust and consist of the Junior Mathematical Olympiad, the European Kangaroo and the International Intermediate Invitational Mathematical Challenges. There is also a summer school for around 50 students in Years 10 and 11 (as defined in England and Wales and their equivalents in Scotland and Northern Ireland).

The British Mathematical Olympiad Committee, now a subtrust of the UKMT, organises two rounds of the British Mathematical Olympiad. Round 1 usually involves about 800 students who have distinguished themselves in the Senior Mathematical Challenge and Round 2 about a hundred. From the latter, twenty are invited to a training weekend at Trinity College, Cambridge. The UK team is then selected and trained for the annual IMO which usually takes place in July. The IMO was held in Argentina in 1997, Taiwan in 1998 and Romania in 1999. The BMOC also sponsors Regional Circles in locations around the country.

The bulk of the administration for all the larger competitions is undertaken by the UK Maths Challenges Office at the University of Leeds by Ms Heather Macklin and Miss Jenny Gill. In the summer of 2000, Angela Gould was appointed Executive Director.

Structure and Membership of the Trust

The governing body of the Trust is its Council. There are presently fifteen members: Dr Peter Neumann (Chair), Dr Roger Bray (Secretary), Professor John Brindley, Mrs Mary Teresa Fyfe, Mr Howard Groves, Mr Terry Heard, Miss Susie Jameson, Dr Imre Leader, Professor Adam McBride, Mr Dennis Orton (Treasurer), Mr Bill Richardson, Professor Chris Robson, Dr Alan Slomson, Miss Patricia Smart and Mr Robert Smart.

As indicated above, the competitions are organised by four subtrusts.

The members of the British Mathematical Olympiad Committee are Professor Adam McBride (Chair), Dr Tony Gardiner (up to July), Dr Imre Leader, Dr Alan Pears (Treasurer), Ms Patricia Smart (Secretary) and Dr Brian Wilson.

The Senior Challenges Subtrust comprises Mr Bill Richardson (Chair), Mr Dennis Archer, Mr Colin Dixon, and Ms Patricia Smart (Treasurer).

The Junior Olympiad Subtrust comprises Professor Chris Robson (Chair), Mrs Mary Teresa Fyfe (Secretary), Dr Andrew Jobbings and Mr Bill Richardson.

The members of the Junior and Intermediate Challenges Subtrust are Professor John Brindley (Chair), Mr Howard Groves, Dr Andrew Jobbings, Professor Chris Robson and Dr Alan Slomson (Secretary and Treasurer).

The members of the Trust comprise past and present members of Council, i.e. the above members of Council together with Dr Tony Gardiner, Mr Peter Thomas and Professor James Wiegold.

An outline of the events

The UKMT provides three Challenge papers and each of these has some sort of follow-up. A brief description of these is given here with much fuller information later in the book.

Junior competitions

The UK Junior Mathematical Challenge is a one hour, 25 question, multiple choice paper for pupils up to:

Y8 in England and Wales;

S2 in Scotland, and

Y9 in Northern Ireland.

Pupils enter their personal details and answers on a special answer sheet for machine reading. The questions are set so that the first 15 should be accessible to all participants whereas the remaining 10 are more testing.

Five marks are awarded for each correct answer to the first 15 questions and six marks are awarded for each correct answer to the rest. Each incorrect answer to questions 16–20 loses 1 mark and each incorrect answer to Questions 21–25 loses 2 marks. Penalty marking is used to discourage guessing.

Certificates are awarded on a proportional basis:– Gold about 6%, Silver about 14% and Bronze about 20% of all entrants. Each centre also receives one 'Best in School Certificate'.

The Junior Mathematical Olympiad is the follow-up competition to the JMC. It is normally held on the second Tuesday in June and between 700 and 1000 high scorers in the JMC are invited to take part. It is a two hour paper which has two sections. Section A contains ten questions and pupils are required to give the answer only. Section B contains six questions for which full written answers are required. It is made clear to candidates that they are not expected to complete all of Section B and that little credit will be given to fragmentary answers. Gold, silver and bronze medals are awarded to very good candidates. In 2000 a total of 29 medals was awarded. As started in 1999, all candidates who sat the paper received a certificate; the top 25% got Certificates of Distinction and the others Certificates of Participation. In addition, the top 50 students were given a book.

Intermediate competitions

The UK Intermediate Mathematical Challenge is organised in a very similar way to the Junior Challenge. One difference is that the age range goes up to Y11, S4 and Y12 in England and Wales, Scotland and Northern Ireland respectively. The other difference is the timing; the IMC is held on the first Thursday in February. All other arrangements are as in the JMC.

The follow-up situation is much more involved than in the Junior Section. High-performing candidates may be eligible for one (or possibly more) tests, depending on their school stage.

About 1000 pupils in Y9 (E & W), S2 (Scot.), Y10 (N.I.) are invited to sit the European Kangaroo which is a one hour, 25 question, multiple choice paper. The last ten questions are more testing than the first fifteen and correct answers gain six marks as opposed to five. (Penalty marking is not applied.) In 2000 the top 25% gained Certificates of Distinction and the others Certificates of Participation. In addition, the top 50 students were given a book. The event took place on Thursday 16th March in 2000.

About 600 pupils in Y10 (E & W), S3 (Scot.), Y11 (N.I.) are invited to sit the International Invitational Intermediate Mathematics Challenge (Grade 10). A similar number in Y11 (E & W), S4 (Scot.), Y12 (N.I.) are invited to sit the International Invitational Intermediate Mathematics Challenge (Grade 11). Both of these originate in Canada and each of the two-hour papers contains five questions to which full written solutions are required. Prizes and certificates are awarded in a similar way to the Kangaroo. One extra feature of the IIIMC papers is that they provide evidence which is used to invite 50 pupils to a five-day summer school early in July. In 2000, the IIIMC papers were taken on Thursday 4th May.

Senior competitions

In 1999 the UK Senior Mathematical Challenge was held on Tuesday 9th November. Like the other Challenges, it is a 25 question, multiple choice paper. However, it lasts 1½ hours and is marked in schools; only a summary of the marks is sent to the organisers to provide information for cut-off scores. Certificates (including Best in School) are awarded as with the other Challenges. The follow-up competitions are organised by the British Mathematical Olympiad Committee.

The first is BMO1, traditionally held on the second Wednesday in January. About 800 are usually invited to take part. The paper lasts 3½ hours and contains five questions to which full written solutions are required.

About 100 high scorers are then invited to sit BMO2, usually on the last Thursday in February, which again lasts 3½ hours but contains four, very demanding, questions.

The results of BMO2 are used to select a group of 20 students to attend a Training Session at Trinity College, Cambridge at Easter. As well as being taught more mathematics and trying numerous challenging problems, this group sits a 4½ hour 'mock' Olympiad paper. On the basis of this and all other relevant information, a group of eight is selected to take part in correspondence courses and assignments which eventually produce the UK Olympiad Team of six to go forward to the International Mathematical Olympiad in July.

The Junior Mathematical Challenge and Olympiad

The Junior Mathematical Challenge was held on Tuesday 28th March 2000, much earlier than normal as Easter was very late. Approximately 142,500 pupils took part. Around 900 were invited to take part in the Junior Mathematical Olympiad which was held on Tuesday 6th June. In the following pages, we shall look at the question paper and solutions leaflet for both events.

We start with the Challenge paper, the front of which is shown below in a slightly reduced format.

UK JUNIOR MATHEMATICAL CHALLENGE

TUESDAY 28TH MARCH 2000

Organised by the **United Kingdom Mathematics Trust from the School of Mathematics, University of Leeds**

RULES AND GUIDELINES (to be read before starting)

1. Do not open the paper until the Invigilator tells you to do so.
2. Time allowed: **1 hour**.
   No answers, or personal details, may be entered after the allowed hour is over.
3. The use of rough paper is allowed; **calculators** and measuring instruments are **forbidden**.
4. Candidates in England and Wales must be in School Year 8 or below.
   Candidates in Scotland must be in S2 or below.
   Candidates in Northern Ireland must be in School Year 9 or below.
5. **Use B or HB pencil only**. Mark *at most one* of the options A, B, C, D, E on the Answer Sheet for each question. Do not mark more than one option.
6. *Do not expect to finish the whole paper in 1 hour.* Concentrate first on Questions 1-15. When you have checked your answers to these, have a go at some of the later questions.
7. Five marks are awarded for each correct answer to Questions 1-15.
   Six marks are awarded for each correct answer to Questions 16-25.
   Each incorrect answer to Questions 16-20 loses 1 mark.
   Each incorrect answer to Questions 21-25 loses 2 marks.
8. Your Answer Sheet will be read only by a *dumb machine*. **Do not write or doodle on the sheet except to mark your chosen options**. The machine 'sees' all black pencil markings even if they are in the wrong places. If you mark the sheet in the wrong place, or leave bits of rubber stuck to the page, the machine will 'see' a mark and interpret this mark in its own way.
9. The questions on this paper challenge you to **think**, not to guess. You get more marks, and more satisfaction, by doing one question carefully than by guessing lots of answers. The UK JMC is about solving interesting problems, not about lucky guessing.

1. What is half of 999?

A 444½ B 449½ C 454½ D 494½ E 499½

2. Sir Isaac Newton, the English mathematician, physicist and discoverer of the laws of gravity, was born in Woolsthorpe, Lincolnshire in 1642, the same year that Galileo, the Italian scientist, died.
How many years ago was that?

A 351 B 358 C 368 D 424 E 442

3. What is the value of x?

A 22 B 28 C 108 D 130 E 208

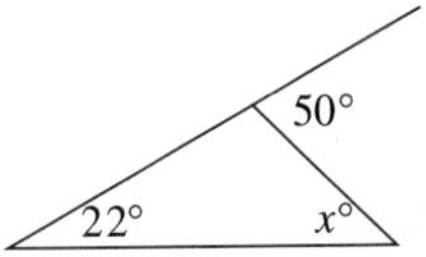

50°
22°
$x°$

4. Which of the following has the greatest value?

A $(1 \times 2) \times (3 \times 4)$ B $(1 \times 2) + (3 \times 4)$ C $(1 \times 2) \times (3 + 4)$
D $(1 + 2) \times (3 \times 4)$ E $(1 + 2) \times (3 + 4)$

5. Which of the following could be the image of U K M T when seen reflected in a mirror?

A ∩KW⊥ B TMKU C UꓘMT D ∩ꓘW⊥ E ⊥Wꓘ∩

6. A transport company's vans each carry a maximum load of 12 tonnes. A firm needs to deliver 24 crates each weighing 5 tonnes. How many van loads will be needed to do this?

A 9 B 10 C 11 D 12 E 13

7. Today, the sun rose at Greenwich at 6:45 am and will set 12 hours and 44 minutes later. At what time will the sun set at Greenwich today?

A 6:29 pm B 7:09 pm C 7:29 pm D 7:39 pm E 9:29 pm

8. A single piece of string is threaded through five holes in a piece of card. One side of the card is shown in the diagram on the right. Which of the diagrams below could *not* represent the pattern of the string on the reverse side?

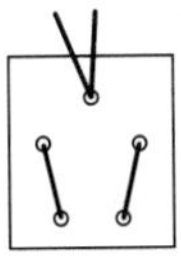

A B C D E

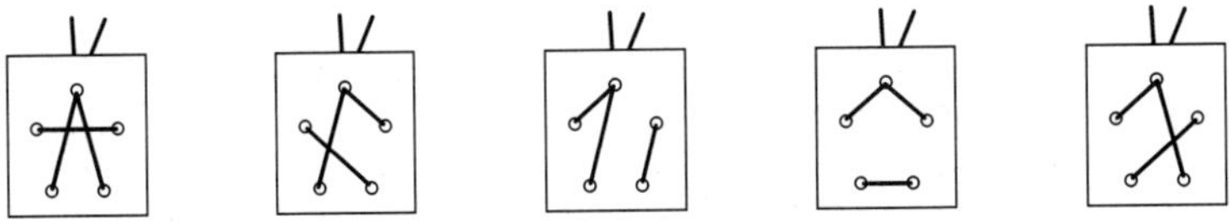

9. Three-quarters of the junior members of a tennis club are boys and the rest are girls. What is the ratio of boys to girls among these members?

A 3 : 4 B 4 : 3 C 3 : 7 D 4 : 7 E 3 : 1

10. Each Junior Mathematical Challenge answer sheet weighs 6 grams. If 140 000 pupils enter the Challenge this year, what will be the total weight of all their answer sheets?

A 84 kg B 840 kg C 8 400 kg D 84 000 kg E 840 000 kg

11. The digits of this year, 2000 A.D., add up to 2. In how many *other* years since 1 A.D. has this happened?

A 3 B 6 C 8 D 9 E 10

12. Four rectangular paper strips, each measuring 10 cm by 1 cm, are laid flat on a table. Each strip is at right angles to two of the other strips as shown.

What is the area of the table covered by the strips?

A 30 cm^2 B 32 cm^2 C 34 cm^2 D 36 cm^2 E 38 cm^2

13. 48% of the pupils at a certain school are girls. 25% of the girls and 50% of the boys at this school travel to school by bus. What percentage of the whole school travel by bus?

A 37% B 38% C 62% D 73% E 75%

14. The DISPUTOR is similar to a calculator, but it behaves a little oddly. When you type in a number, the DISPUTOR doubles the number, then reverses the digits of this result, then adds 2 and displays the final result. I type in a whole number between 10 and 99 inclusive. Which of the following could be the final result displayed?

A 39 B 41 C 42 D 43 E 45

15. Dilly is 7 years younger than Dally. In 4 years time she will be half Dally's age. What is the sum of their ages now?

A 13 B 15 C 17 D 19 E 21

16. A book has 256 pages with, on average, 33 lines on each page and 9 words on each line. Which of the following is the best approximation to the number of words in the book?

A 64 000 B 68 000 C 72 000 D 76 000 E 80 000

17. The first and third digits of the five-digit number $d6d41$ are the same. If the number is exactly divisible by 9, what is the sum of its five digits?

A 18 B 23 C 25 D 27 E 30

18. A circle is added to the grid alongside. What is the largest number of dots that the circle can pass through?

A 4 B 6 C 8 D 10 E 12

19. The numbers $\frac{1}{2}$, x, y, $\frac{3}{4}$ are in increasing order of size. The differences between successive numbers in this list are all the same. What is the value of y?

A $\frac{3}{8}$ B $\frac{2}{3}$ C $\frac{7}{12}$ D $\frac{5}{6}$ E $\frac{5}{8}$

20. Despite his name, Mr. Bean likes to eat lots of fruit. He finds that four apples and two oranges cost £1.54 and that two oranges and four bananas cost £ 1.70. How much would he have to pay if he bought one apple, one orange and one banana?

A 77p B 78p C 79p D 80p E 81p

21. Tick's watch runs 2 minutes per hour too slow. Tock's watch runs 1 minute per hour too fast. They set them to the same time at noon on Sunday. The next time they met, one of the watches was one hour ahead of the other. What was the earliest time this could have been?

A 8am on Monday B 7:20 pm on Monday C 4am on Tuesday

D midnight on Wednesday E 10 pm on Saturday

22. Four identical blocks of wood are placed touching a table in the positions shown in this side-on view.
How high is the table?

A 84cm B 87cm C 90cm D 93cm E 96cm

Table
84 cm
96 cm
Floor

23. A certain number has exactly eight factors including 1 and itself. Two of its factors are 21 and 35. What is the number?

A 105 B 210 C 420 D 525 E 735

24. The six cards shown display the number 632579. One "turn" consists of exchanging the positions of two adjacent cards so, for instance, after one "turn" the cards could show 632759. Starting from the original 632579, what is the least number of "turns" required so that the cards display a number which is divisible by 4?

6 3 2 5 7 9

A 2 B 3 C 4 D 5 E 6

25. In a magic square each row, each column and both main diagonals have the same total. What number should replace x in this partially completed magic square?

13		
5		15
x		

A 4 B 9 C 10 D 12

E more information needed

The JMC solutions

As in 1999, there was a solutions leaflet.

UK JUNIOR MATHEMATICAL CHALLENGE

TUESDAY 28TH MARCH 2000

Organised by the **United Kingdom Mathematics Trust from the School of Mathematics, University of Leeds**

SOLUTIONS LEAFLET

This solutions leaflet for the JMC is sent in the hope that it might provide all concerned with some alternative solutions to the ones they have obtained. It is not intended to be definitive. The organisers would be very pleased to receive alternatives created by candidates.

1. **E** Half of 999 = ½ (1000 −1) = 500 − ½ = 499½.

2. **B** 2000 − 1642 = 358.

3. **B** $x = 50 - 22$. This method uses the theorem:
An exterior angle of a triangle is equal to the sum of the two interior and opposite angles.

4. **D** The values are A 24 B 14 C 14 D 36 E 21.

5. **A** If UKMT is reflected in a horizontal mirror line directly below it then image A appears.

6. **D** A van load can take only two crates and hence twelve van loads will be required.

7. **C** 12 hours and 15 minutes from 6:45 am will take the time to 7:00 pm and there will then be 29 minutes of the 12 hours and 44 minutes remaining.

8. **C** In C there is no connection between the pair of holes on the left of the card and the pair of holes on the right. There must, however, be such a connection for the display on the front of the card to be as shown.

9. **E** If three-quarters of the members are boys then one-quarter are girls and the number of boys is three times the number of girls.

10. **B** 140 000 × 6 g = 840 000 g = 840 kg. *Almost one tonne!*

11. **D** The relevant years are 2, 11, 20, 101, 110, 200, 1001, 1010 and 1100.

12. **D** The total area of the four strips is 40 cm^2, but there are four squares, each of area 1 cm^2, where two strips overlap. Hence the area covered is $(40 - 4)\ \text{cm}^2$.

13. **B** The girls who travel by bus make up 12% (¼ of 48%) of the whole school and the corresponding figure for boys is 26% (½ of 52%). Hence 38% of the whole school travel by bus.

14. **E** In reverse, the stages which lead to 45 are: $45 \leftarrow 43 \leftarrow 34 \leftarrow 17$.

15. **A** Let Dilly's age be x. Then Dally is $x + 7$. In four years time Dilly will be $x + 4$ and Dally will be $x + 11$. Therefore $x + 11 = 2(x + 4)$ and hence $x = 3$. Dilly is 3, Dally is 10 and the sum of their ages is 13.

16. **D** The number of words $= 256 \times 33 \times 9 \approx 250 \times 300 = 75\ 000$. The best estimate, therefore, is 76 000.

17. **D** If a number is divisible by 9, then the sum of its digits must also be a multiple of 9. The sum of the digits of $d6d41$ is $2d + 11$, which must be an odd number between 13 and 29 inclusive. The only odd multiple of 9 in this interval is 27.

18. **C** Let the distance between adjacent dots be one unit.
Then a circle of radius $\sqrt{5}$ units whose centre is at the centre of the grid passes through eight dots, as shown.

19. **B** The difference between successive numbers in the list is $\frac{1}{3}\left(\frac{3}{4} - \frac{1}{2}\right) = \frac{1}{12}$.
Therefore $y = \frac{3}{4} - \frac{1}{12} = \frac{2}{3}$.

20. E The total cost of four apples, four oranges and four bananas is £1.54 + £1.70 = £3.24. Hence the amount Mr. Bean would pay for one apple, one orange and one banana is £3.24 ÷ 4 = 81p.

21. A Each hour, Tock 's watch gains three minutes on Tick's watch. It will, therefore, take 20 hours before it is one hour ahead of it.

22. C Let the width and height of each block of wood be x and y respectively and the height of the table be h. Then: $h + x = y + 84$ and $h + y = x + 96$.

Add these two equations: $h + x + h + y = y + x + 84 + 96$

Subtracting $(x + y)$ from both sides: $2h = 180$

$\therefore \quad h = 90$

23. A If 21 and 35 are factors of the number, then 3, 5 and 7 must all be included amongst its prime factors. This means that the required number must be a multiple of 105 and, as the complete list of factors of 105 is 1, 3, 5, 7, 15, 21, 35 and 105, the answer is 105.

(A number which is the product of three different prime numbers must have exactly eight factors. If p, q and r are all different prime numbers then the factors of pqr are 1, p, q, r, pq, pr, qr and pqr itself.)

24. B Exchanging the '2' and the '5', then the '2' and the '7' and, finally, the '2' and the '9' gives a display of 635792, which is a multiple of 4.

(A whole number is a multiple of 4 if, and only if, the number formed by its last two digits, in this case 92, is a multiple of 4. Can you prove this?)

25. D Comparing the leading diagonal and the second row:
$13 + c + e = 5 + c + 15$ and hence $e = 7$.
Comparing the top row and the third column:
$13 + a + b = b + 15 + e$ and hence $a = e + 2 = 9$.
Comparing the second row and the second column:
$a + c + d = 5 + c + 15$ and hence $d = 20 - a = 11$.

13	a	b
5	c	15
x	d	e

We now have:

13	9	b
5	c	15
x	11	7

Let the 'magic' total be T.

Then: $T = 22 + b$
$T = 20 + c$
$T = 18 + x$
$T = b + c + x$.

Adding the first three of these equations:
$3T = 60 + b + c + x$
Hence: $3T = 60 + T$
Thus: $T = 30$.
Therefore $x = 30 - 18 = 12$.

(Notice that the magic total, in this case 30, is three times the number in the middle of the magic square, in this case 10. Can you prove that this is always the case in a 3 × 3 magic square?)

The JMC answers

The table below shows the proportion of pupils' choices. The correct answer is shown in bold.

Q	A	B	C	D	E	Blank
1	8	8	8	4	**71**	1
2	1	**88**	8	1	2	1
3	3	**64**	25	3	2	2
4	13	1	1	**83**	2	1
5	**47**	16	17	9	9	1
6	3	65	3	**25**	2	1
7	10	2	**79**	5	2	1
8	11	7	**54**	15	8	3
9	24	8	2	1	**63**	2
10	4	**34**	7	7	46	1
11	35	23	13	**14**	11	3
12	14	11	12	**47**	7	8
13	2	**32**	6	5	52	2
14	5	5	35	11	**37**	6
15	**36**	12	12	10	21	9
16	8	6	7	**51**	10	18
17	12	7	6	**39**	3	33
18	18	4	**12**	8	31	27
19	4	**32**	7	5	10	43
20	5	6	7	7	**17**	57
21	**25**	9	6	11	3	47
22	6	8	**25**	7	5	48
23	**16**	13	7	4	11	49
24	7	**20**	9	7	4	52
25	4	5	4	**12**	38	37

JMC 2000: Some comments on the pupils' choices of answers

The Problems Group succeeded in its aim of setting a more accessible paper and the average score rose from 39.8 in 1999 to 48.6 in 2000. Nonetheless there were some surprises.

Question 1 was meant to be a very easy starter, but only 71% of the entrants got it right. We hoped that pupils would argue that since 999 is 1 less than 1000, a half of 999 is a half less than a half of 1000, a line of argument which rapidly leads to the correct answer. We suspect that the use of calculators means that children are no longer encouraged to think in this sort of way, and that faced with having to divide 999 by 2 without a calculator, the 29% who got the question wrong either made a slip or a wild guess.

We know that there is a lot of time pressure in doing the Challenge, but it was disappointing that 65% of the entrants fell into the obvious trap in Question 6 because they did not stop to think. Shortage of time no doubt encourages children to think that if some of the numbers in the question can be combined to give one of the suggested answers, then this must be the right answer, without the need to think carefully which is the relevant sum. For example, Question 9 begins with ‘three-quarters’ and 24% of the entrants chose the answer 3:4. In Question 13, 25% plus 50% gives 75% which 52% of the pupils chose as the correct answer, even though it should be clear from the given information that the correct answer must be lower than 50%.

‘More information is needed’ is always a popular answer, and so it was not surprising that 38% of the entrants chose this answer to Question 25, and that a higher proportion of the entrants than usual gave an answer to this final question which is always a hard one. In this case it turns out that, contrary to what you might first think, the information given is sufficient to determine that the value of x is 12.

The profile of marks obtained is shown below.

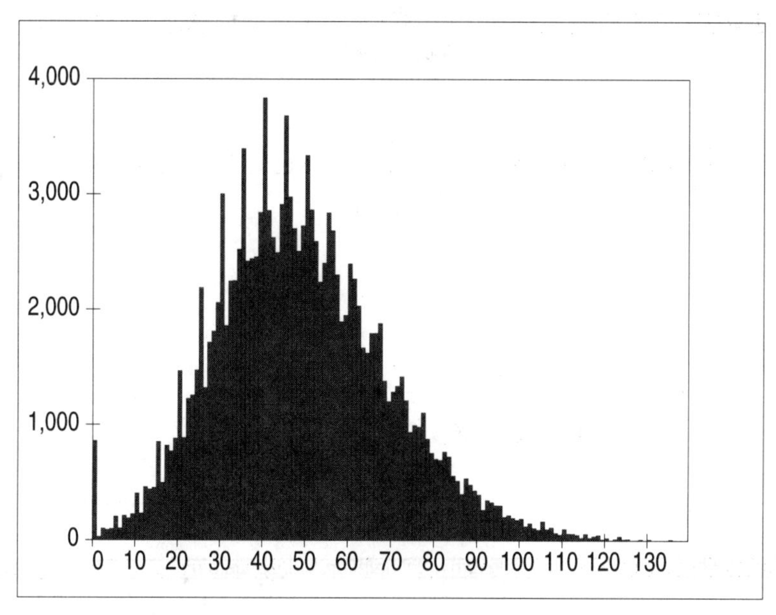

Bar chart showing the actual frequencies in the 2000 JMC

On the basis of the standard proportions used by the UKMT, the cut-off marks were set at

GOLD – 82 or over SILVER – 66 to 81 BRONZE – 52 to 65

A sample of one of the certificates is shown on the next page.

The Junior Mathematical Olympiad is the follow-up competition to the Challenge. It was decided that candidates who obtained a JMC score of 106 or over were eligible to take part in the JMO. This resulted in some 942 candidates being invited.

UK JUNIOR MATHEMATICAL CHALLENGE
2000

. . . .

of

. . . .

received a

GOLD CERTIFICATE

Peter Neumann

Chairman, United Kingdom Mathematics Trust

THE UNITED KINGDOM JUNIOR MATHEMATICAL CHALLENGE

The UK JMC encourages mathematical reasoning, precision of thought, and fluency in using basic mathematical techniques to solve non-standard problems. It is targeted at the top third of pupils in English School Years 7-8 (S1 and S2 in Scotland and Years 8-9 in Northern Ireland).

The problems on the UK JMC are designed to make students think, and sometimes smile. Most are accessible to younger students, yet still challenge those with more experience; they are also meant to be memorable and enjoyable.

Mathematics controls more aspects of the modern world than most people realise – from CDs, cash machines, telecommunications and airline booking systems to production processes in engineering, efficient distribution and stock-holding, investment strategies and 'whispering' jet engines. The scientific and industrial revolutions flowed from the realisation that mathematics was both the language of nature, and also a way of analysing – and hence controlling – our environment. In the last fifty years old and new applications of mathematical ideas have transformed the way we live.

All these developments depend on mathematical thinking – a mode of thought whose essential style is far more permanent than the wave of technological change which has made so many recent changes possible. The problems on the UK JMC reflect this style – a style which pervades all mathematics – by challenging students to think clearly about simple, yet unfamiliar problems.

The UK JMC grew out of a national challenge first run in 1988. In 1999 over 141,000 entries were received, an increase of 8% over 1998, and over 2,000 schools took part. Certificates were awarded to the highest scoring 40% of candidates (6% Gold, 13% Silver, 21% Bronze).

There is an Intermediate version for those in Years 9 to 11, and a Senior version for those in Years 12 and 13. All three events are organised by the United Kingdom Mathematics Trust and are administered from the School of Mathematics at the University of Leeds.

The Junior Mathematical Olympiad

UK Junior Mathematical Olympiad 2000

Organised by The United Kingdom Mathematics Trust

Tuesday 6th June 2000

RULES AND GUIDELINES : READ THESE INSTRUCTIONS CAREFULLY BEFORE STARTING

1. Time allowed: 2 hours.
2. **The use of calculators *and measuring instruments* is forbidden.**
3. All candidates must be in *School Year 8 or below* (English and Wales), *S2 or below* (Scotland), *School Year 9 or below* (Northern Ireland).
4. For questions in Section A *only the answer is required.* Enter each answer neatly in the relevant box on the Front Sheet. Do not hand in rough work.

   For questions in Section B you must give *full written solutions*, including clear mathematical explanations as to why your method is correct.

   Solutions must be written neatly on A4 paper. Sheets must be STAPLED together in the top left corner with the Front Sheet on top.

   Do not hand in rough work.
5. Questions A 1-10 are relatively short questions. Try to complete Section A within the first hour so as to allow at least one hour for Section B.
6. Questions B1-B6 are longer questions requiring *full written solutions*.
   This means that each answer must be accompanied by clear explanations and proofs.
   Work in rough first, then set out your final solution with clear explanations of each step.
7. These problems are meant to be hard! Do not hurry. Try the earlier questions in each section first (they tend to be easier). Try to finish whole questions even if you can't do many. A good candidate will have done most of Section A and given solutions to at least two questions in Section B.
8. Numerical answers must be FULLY SIMPLIFIED, and EXACT using symbols like π, fractions, or square roots if appropriate, but NOT decimal approximations.

DO NOT OPEN THE PAPER UNTIL INSTRUCTED BY THE INVIGILATOR TO DO SO!

Section A

A1 What is the value of $2000 + 1999 \times 2000$?

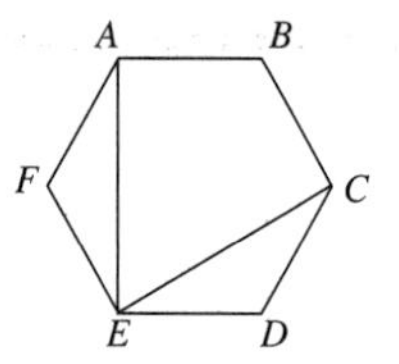

A2 $ABCDEF$ is a regular hexagon.
What fraction of the area of the hexagon is the area of the kite $ABCE$?

A3 Estimate, correct to the nearest mm, the side of a square of area $0{\cdot}5\ \text{cm}^2$.

A4 What is 20% of 30% of 40% of £50?

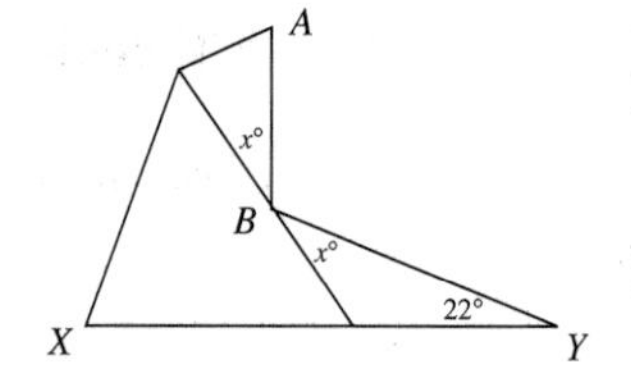

A5 The diagram shows part of a mosaic of tiles.
AB is vertical and XY is horizontal.

What is the value of x?

A6 The hour hand on the Mad Hatter's watch moves at the correct speed, but the minute hand moves one and a half times as fast as it should. Yesterday, it showed the correct time at 3 p.m. When did it next show the correct time?

A7 Six pupils have, between them, won three gold medals, two silver medals and a bronze medal in a painting competition. Unfortunately, their teacher, Mr. Turner, has lost all record of which medals should go to which pupils, so he allocates them by drawing names out of a hat. The first three names drawn receive the gold medals, the next two drawn have the silver medals and the bronze medal goes to the remaining pupil.
In how many different ways can the medals be allocated by this method?

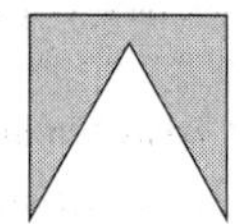

A8 An equilateral triangle is cut out of a square of side 2 cm, as shown.

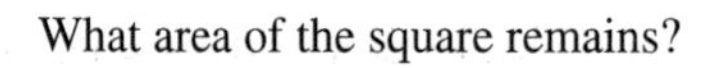

What area of the square remains?

A9 The machine which prints photographs at *Snippysnaps* runs for the same time every day. It prints colour photographs at a constant rate and monochrome (black and white) photographs at a different constant rate.
On Monday, the machine printed 2100 colour photographs and 2450 monochrome photographs.
On Tuesday it printed 2800 colour photographs and 1400 monochrome photographs.
On Wednesday, it printed only monochrome photographs.
How many of these were there?

A10 It takes four gardeners four hours to dig four circular flower beds, each of diameter four metres. How long will it take six gardeners to dig six circular flower beds, each of diameter six metres?

Section B

Your solutions to Section B will have a major effect on the JMO results. Concentrate on one or two questions first and then **write out full solutions** (not just brief 'answers').

B1 Kate has 90 identical building blocks. She uses all of the blocks to build this four-step 'staircase' in which each step, apart from the top one, is the same length.

(i) Show that there are exactly two different ways in which it is possible to use all 90 blocks to build a six-step 'staircase'.

(ii) Explain fully why it is impossible to use all 90 blocks to build a seven-step 'staircase'.

B2 A crossnumber puzzle is like a crossword puzzle – except that the answers are numbers instead of words and each square contains one single digit. None of the answers starts with the digit 0.

How many solutions are there to this crossnumber?

(You must use logic, not guesswork.)

Across
1. Square
3. Square
4. Square

Down
1. Cube
2. Square
3. Cube times square

B3 The diagram shows an equilateral triangle inside a rhombus. The sides of the rhombus are equal in length to the sides of the triangle.

What is the value of x?

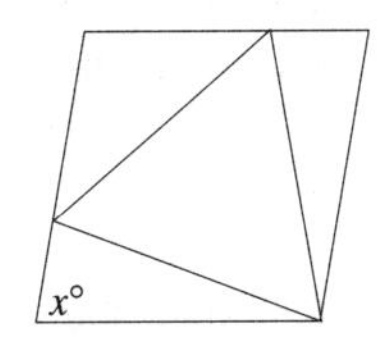

B4 How many different solutions are there to the letter sum on the right?

Different letters stand for different digits, and no number begins with a zero.

$$\begin{array}{r} \text{J M C} \\ +\,\text{J M O} \\ \hline \text{S U M S} \end{array}$$

B5 (i) Explain why the sum of three consecutive integers is always divisible by 3.

(ii) Is it true that the sum of four consecutive integers is always divisible by 4?

(iii) For which k is it true that the sum of k consecutive integers is always divisible by k?

B6 X and Y play a game in which X starts by choosing a number, which must be either 1 or 2.

Y then adds either 1 or 2 and states the total of the two numbers chosen so far. X does likewise, adding either 1 or 2 and stating the total, and so on. The winner is first player to make the total reach (or exceed) 20.

(i) Explain how X can always win.

(ii) The game is now modified so that at each stage the number chosen must be 1 or 2 or 4. Which of X or Y can now always win and how?

UK Junior Mathematical Olympiad 2000 Solutions

A1
4000000

$2000 + 1999 \times 2000 = (1 + 1999) \times 2000 = 2000 \times 2000 = 4\ 000\ 000.$

A2
$\mathbf{\frac{2}{3}}$

The regular hexagon may be divided up into six congruent triangles, as shown. The kite $ABCE$ is made up of four of these.

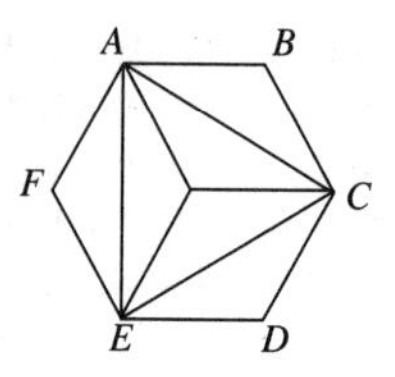

A3
7 mm

The area of the square = 0.5 cm^2 = 50 mm^2.
Therefore the side of the square = $\sqrt{50}$ mm = 7 mm (to the nearest mm).

A4
£1.20

20% of 30% of 40% of £50 = $\frac{2}{10} \times \frac{3}{10} \times \frac{4}{10} \times$ £50 = $\frac{24}{1000} \times$ £50 = £$\frac{1200}{1000}$ = £1.20.

A5
34

Produce AB so that it meets XY at C. Then, since AB is vertical and XY horizontal, $\angle ACY$ is a right angle. Now $\angle CBD = \angle EBA = x°$ (vertically opposite angles).
Therefore, in ΔCBY :
$22 + x + x + 90 = 180$
i.e. $2x = 68$ i.e. $x = 34$.

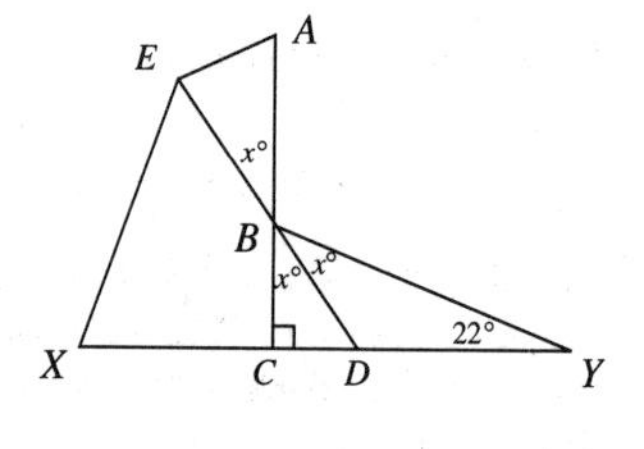

A6
5 p.m.

The Mad Hatter's watch next showed the correct time when the minute hand had gained exactly one hour. This took two hours to do since the minute hand makes three complete revolutions every two hours. Hence it next showed the correct time at 5 p.m.

A7
60

The number of different ways in which the medals may be allocated is clearly the same as it would be if Mr. Turner first chose one pupil for the Bronze medal and then two pupils for the Silver medals. The three remaining pupils would then receive the Gold medals.

There are six possible ways to allocate the Bronze medal. For each choice of the Bronze medallist, the two Silver medallists have to be chosen from the 5 remaining pupils: here there are 5 ways to choose the first Silver medallist A, and for each such choice there are four ways to choose the second Silver medallist B. So there are 5×4 ways of choosing the Silver medallists "in order". However, AB and BA are then counted separately, so, once the Bronze medallist has been chosen, there are $\frac{5 \times 4}{2} = 10$ unordered ways to choose the two Silver medallists. The Gold medallists are then determined. Hence there are 6×10 ways to allocate the medals to the six pupils.

A8 Let the height of the equilateral triangle be h cm.

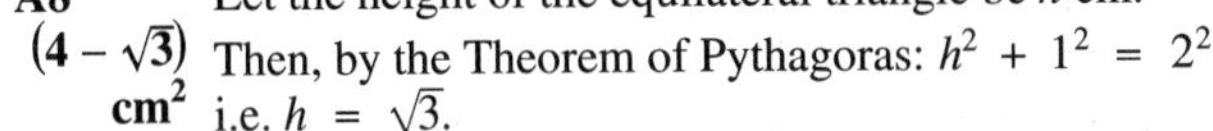

$(4-\sqrt{3})$ cm^2 Then, by the Theorem of Pythagoras: $h^2 + 1^2 = 2^2$ i.e. $h = \sqrt{3}$.

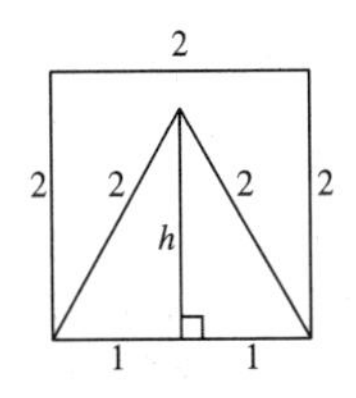

Hence area of triangle $= \frac{1}{2} \times 2 \times \sqrt{3}\,\text{cm}^2 = \sqrt{3}\,\text{cm}^2$.

Therefore, the area of the square which remains is $(4 - \sqrt{3})\ \text{cm}^2$.

A9
5600 On Tuesday, the machine printed 700 more colour photographs than on Monday, but 1050 fewer monochrome photographs. In the time it took to print 2800 colour photographs, therefore, the number of monochrome photographs it could have printed is $4 \times 1050 = 4200$. Hence the number of monochrome photographs the machine printed on Wednesday is $4200 + 1400 = 5600$.

A10
9 hours One gardener takes four hours to dig one flower bed of diameter four metres. Now the time taken to dig a circular flower bed is directly proportional to its area and hence directly proportional to the square of its diameter.

Thus: $\dfrac{\text{time taken to dig bed of diameter six metres}}{\text{time taken to dig bed of diameter four metres}} = \left(\dfrac{6}{4}\right)^2 = \left(\dfrac{3}{2}\right)^2 = \dfrac{9}{4}$.

Hence the time taken for one gardener to dig a bed of diameter six metres is $\frac{9}{4} \times 4$ hours = 9 hours.

The time taken for six gardeners to dig six flower beds is the same as the time it takes one gardener to dig one bed of the same size and hence the answer is 9 hours.

Section B

B1 (i) Let the number of blocks on the top step be t and the length of each step be l. Then:

Number of blocks in a six-step 'staircase'

$= t + (t + l) + (t + 2l) + (t + 3l) + (t + 4l) + (t + 5l) = 6t + 15l.$

Hence $6t + 15l = 90$ i.e. $2t + 5l = 30$.

We deduce that l must be even (*why?*) and also that $l < 6$ (since $t \leqslant 0$ when $l \geqslant 6$).

When $l = 2$: $2t + 10 = 30$ i.e. $t = 10$.

When $l = 4$: $2t + 20 = 30$ i.e. $t = 5$.

Thus there are exactly two different ways of building six-step 'staircases' using all the blocks: 10, 12, 14, 16, 18, 20 and 5, 9, 13, 17, 21, 25.

(ii) The number of blocks in a seven-step 'staircase' $= 6t + 15l + (t + 6l) = 7t + 21l = 7(t + 3l)$.

Thus we would require $7(t + 3l) = 90$. However, 90 is not a multiple of 7 and therefore a seven-step 'staircase' using all 90 blocks is not possible.

B2 A sensible place to start would be to try to find 1 Down since there are only five three-digit cubes, namely 125, 216, 343, 512 and 729.

We may rule out 343 and 512 because 4 Across is a square and so cannot end in either '3' or '2'. Also, 729 may be ruled out since 1 Across cannot begin with '7'.

If 1 Down is 216, then 1 Across must be 25. However, 2 Down cannot start with '5' and hence 216 can also be eliminated.

Any solution, therefore, must have 125 as 1 Down. This means that 1 Across and 2 Down must be 16 and 64 respectively: there are no other correct possibilities. The only two-digit square ending in '5' is 25 and hence 4 across must be 25.

We note that 3 Across is a three-digit square ending in 24 and the only possibility is $18^2 = 324$.

We have now filled in all the squares, with there being no other correct possibilities, but we must check that 3 Down is correct. As $32 = 8 \times 4 = 2^3 \times 2^2$, it is indeed a cube times a square and therefore we deduce that there is exactly one solution to the crossnumber.

B3 ΔDFC is isosceles ($CF = CD$).

Hence $\angle DFC = \angle FDC = x°$.

Hence $\angle FCD = (180 - 2x)°$ (angle sum of triangle).

Now $\angle EBC = \angle FDC = x°$ (opposite angles of a parallelogram) and ΔEBC is isosceles ($CE = CB$).

Hence $\angle BEC = x°$ and $\angle ECB = (180 - 2x)°$.

Lines AD and BC are parallel and hence $\angle ADC + \angle BCD = 180°$.

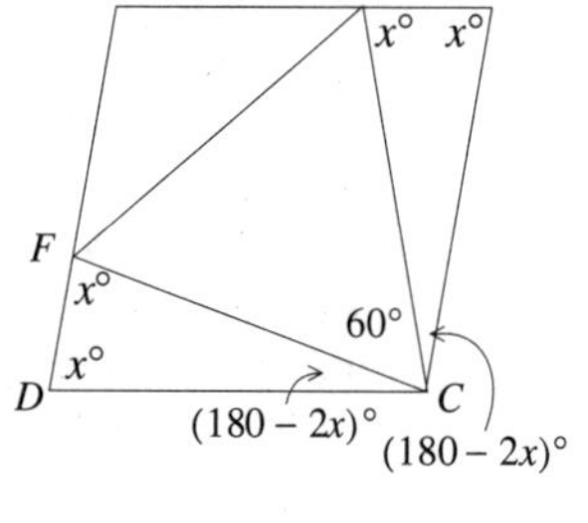

Therefore: $x + 2(180 - 2x) + 60 = 180$ i.e. $420 - 3x = 180$ i.e. $3x = 240$ so $x = 80$.

B4 Each of the two three-digits numbers is less than 1000. Their sum, therefore, is less than 2000 and we can deduce that $S = 1$. We note now that C + O must equal either 1 or 11. However, since neither C nor O can have the same value as S (i.e. 1), their sum cannot be 1 and hence C + O = 11. There is a carry of 1 into the tens column, therefore, and, since 1 + M + M cannot equal M, we may deduce that 1 + M + M = M + 10 i.e. M = 9.

This means that there is a carry of 1 into the hundreds column and hence 1 + J + J = U + 10.

We must, therefore, find all possible values of C, O, J and U which satisfy C + O = 11 and U = 2J – 9, remembering that none of these four letters may be either 1 or 9.

From the equation U = 2J – 9, we deduce that J $\geqslant$ 5.

If J = 5 then U = 1 (impossible).

If J = 6 then U = 3. Hence C = 4 and O = 7 or vice versa.

If J = 7 then U = 5. Hence C = 3 and O = 8 or vice versa.

If J = 8 then U = 7. Hence C = 5 and O = 6 or vice versa.

Thus there are six different solutions:

694	697	793	798	895	896
697	694	798	793	896	895
1391	1391	1591	1591	1791	1791

B5 (i) Let the smallest of the three integers be s. Then the others are $(s + 1)$ and $(s + 2)$.

Their sum $= s + (s + 1) + (s + 2) = 3s + 3 = 3(s + 1)$ is therefore always divisible by 3.

However, a method which will be more useful for part (iii) of this question is to let the middle number be m. Then the smallest of the three numbers is $(m - 1)$ and the largest $(m + 1)$.

Their sum $= (m - 1) + m + (m + 1) = 3m$ is therefore always divisible by 3.

(ii) No. It is not true. In part (i) where we are required to explain why something is always true, we need to use algebra. However, when we need to show that a statement is not always true, one example (known as a 'counter-example') is sufficient.

Note that $1 + 2 + 3 + 4 = 10$, which is not divisible by 4, and thus we have shown that the sum of four consecutive positive integers is not always divisible by 4.

Note:

We could, if required, show that the sum of four consecutive integers is **never** divisible by 4. Let the smallest of the four integers be s. Then the others are $(s + 1)$, $(s + 2)$ and $(s + 3)$. Their sum $= s + (s + 1) + (s + 2) + (s + 3)$ $= 4s + 6$, which is not divisible by 4.

(iii) A little investigation appears to show that the sum of k integers is divisible by k when k is odd, but not when k is even. However, as in part (i), we must use algebra to prove that this is **always** the case.

Consider the case when k is odd: let $k = 2n + 1$ where n is a positive integer.

Let the middle number of these k numbers be m. Then the smallest of the numbers is $(m - n)$ and the largest $(m + n)$. Let their sum be S. Then:

$$S = (m - n) + (m - (n - 1)) + (m - (n - 2)) + \ldots + (m - 1) + m + (m + 1) + \ldots + (m + (n - 2)) + (m + (n - 1)) + (m + n).$$

Note that $(m - n) + (m + n) = 2m$; $(m - (n - 1)) + (m + (n - 1)) = 2m$; $(m - (n - 2)) + (m + (n - 2)) = 2m$ etc.

There are n pairs of terms which each sum to $2m$ and one remaining term, the middle term, m. Hence: $S = n \times 2m + m = (2n + 1)m = km$.

We deduce that the sum of k consecutive integers is always divisible by k when k is odd.

Consider the case when k is even. Our conjecture is that the sum of k consecutive integers is **not** always divisible by k when k is even. To prove that this is correct we need only to find one counter-example for each value of k.

Let $k = 2n$ where n is a positive integer.

$$\begin{array}{lllllllllllll} \text{Let} & S = & 1 & + & 2 & + & 3 & + & \dots & + & (2n-1) & + & 2n \\ \text{Then} & S = & 2n & + & (2n-1) & + & (2n-2) & + & \dots & + & 2 & + & 1 \end{array}$$

Adding these two equations gives:

$$\begin{aligned} 2S &= (2n+1) + (2n+1) + (2n+1) + \dots + (2n+1) + (2n+1) \\ &= 2n(2n+1). \end{aligned}$$

Therefore $S = n(2n + 1)$, which is an **odd** multiple of n and therefore cannot be a multiple of $2n$. Thus when k is even, the sum of the first k positive integers is not divisible by k and so we deduce that the sum of k consecutive positive integers is not always divisible by k when k is even.

(It is possible to prove that the sum of k consecutive integers is never divisible by k when k is even, but this is not necessary in this case.)

Thus the sum of k consecutive integers is divisible by k if and only if k is odd.

B6 (i) X chooses 2 first. Whatever Y chooses, X can now make the total 5 i.e. if Y chooses 1 then X chooses 2 and vice versa. X repeats this tactic, choosing the number which Y does not choose, thereby ensuring that after four more turns each the totals are 8, 11, 14 and 17 respectively. If Y now chooses 1, making the total 18, X chooses 2 to make the total 20 and win the game. Alternatively, if Y chooses 2, then X chooses 1 and still wins.

(ii) A similar tactic is required for the revised game. The important totals, i.e. the 'winning positions', this time are 3, 6, 9, 12 and 15; a player achieving any of these totals can win the game by best play. However, a player who makes the total 1 or 2 or 4 or 5 or 7 etc. gives the opponent the opportunity of establishing and maintaining a 'winning position'. Since Y can achieve either 3 or 6 after one turn but X cannot, it follows that it is Y who can always win. This is shown in the following table.

	A			B			C		
Original Total	0	0	0	3	3	3	6	6	6
X chooses	1	2	4	1	2	4	1	2	4
Y chooses	2	4	2	2	4	2	2	4	2
New Total	3	6	6	6	9	9	9	12	12
Comment	Go to B	Go to C	Go to C	Go to C	Go to D	Go to D	Go to D	Go to E	Go to E

	D			E			F		
Original Total	9	9	9	12	12	12	15	15	15
X chooses	1	2	4	1	2	4	1	2	4
Y chooses	2	4	2	2	1	4	4	4	1
New Total	12	15	15	15	15	20	20	21	20
Comment	Go to E	Go to F	Go to F	Go to F	Go to F	Y wins	Y wins	Y wins	Y wins

(Other choices are available to Y in some cases e.g. column 2 of B could be:

3 2 1 6 Go to C)

Note that game (i) is equivalent to the game in which players take turns to take either one or two matches from an original pile of 20 matches with the winner being the player who takes the last match.

The marking and results

The pupils' scripts began arriving very rapidly and the marking took place in Leeds the following weekend, 10th and 11th June. Howard Groves led the discussions as to what marks should be given and for what. The marking team was: Terry Heard, Andrew Jobbings, Alex Voice, Alan Slomson, Jenny Ramsden, Dean Bunnell, Gerry Leversha, Professor John Webb (University of Cape Town!) and Mike Moon.

As has been stated, the object of the JMO is for pupils to be *challenged*, possibly in ways they have not been before. Some participants may find all of Section B rather beyond them, but it is hoped that they achieve a degree of satisfaction from Section A. Satisfaction is an important aspect of this level of paper; nevertheless those who do succeed in tackling Section B deserve credit for that and such credit is mainly dependent on getting solutions to questions in Section B which are 'perfect' or very nearly so. The awarding process is somewhat complicated, some might say bizarre. Firstly there are certificates which come in two versions, Participation and Distinction. Everyone is awarded a certificate with about a quarter obtaining a Certificate of Distinction. There were book prizes for the top fifty, the book in question being *Mathematical Challenges III* published by The Scottish Mathematical Council. Finally, there were medals of the traditional Gold, Silver, Bronze varieties.

The paper itself was found to be a little harder than that of 1999. As a consequence, fewer medals were awarded: 4 Gold, 14 Silver and 21 Bronze. A list of these is shown below.

The list below includes all the medal winners in the 2000 JMO. Within each category, the names are in alphabetical order.

GOLD MEDALS

Cong Chen	Leicester Grammar School
Chris Kerr	Downsend School
Martin Orr	Methodist College
Alex Smith	KE VI Five Ways School

SILVER MEDALS

Ellis Bowyer	John Hampden GS
Rikesh Chotai	St Christophers School
Jacob Davis	John Hampden GS
Chris Dorrington	Stamford School
Jack Farchy	St Andrews School
Peter Ford	RGS Worcester
Elena Gordeeva	St Annes Convent School
Nathan Kettle	Hitchin Boys
Richard Lau	KES Birmingham
Stephen Legg	Toot Hill C S
Robert Miles	Newland House School
Jonathan Rees	Milbourne Lodge School
Ran Wei	Olchfa Comprehensive
Lee Zhao	Nottingham H S

BRONZE MEDALS

Steven Bruce	Newcastle Under Lyme
Anna Button	Park View School
Jonathan Cairns	Reading School
Edward Carlsson-Browne	Colchester RGS
James Cleave	Nottingham H S
Charles Franklin	Highgate Junior School
Benjamin Humphries	Kings School Winchester
Philippa Kennedy	St Pauls Girls School
Arjun Mehta	Reddiford School
Iain Monro	Millfield Preparatory
Amy Pang	Haberdashers Askes Girls
Tiffany Ritchie	Monifieth High School
Elizabeth Roberts	Cheadle Hulme School
David Roden	KE VI Camp Hill Boys School
David Sankey	Reading School
Ben Sherlock	Holmewood House
Mark Thompson	Douglas Academy
Thomas Wallace	St Michaels, Barnstable
Ng Yin Ting	Belmont
Fraser Young	The Skinners School
Yao Zhou	Tiffin School

The results and all the extras (books, certificates and medals) were sent away with every chance of reaching schools before pupils departed for their summer holidays.

Readers with good memories may recognise the name of Martin Orr; he won a Gold Medal in 1999 as well!

The Intermediate Mathematical Challenge and its follow-up events

The Intermediate Mathematical Challenge was held on Thursday 3rd February 2000. Approximately 123,500 pupils took part. There were three different follow-up competitions and pupils were invited to the one appropriate to their school year. Pupils in Year 9 (English style) went on to sit the European Kangaroo, those in Years 10 and 11 took the International Invitational Intermediate Mathematics Competition 10 or 11 respectively. We start with the IMC.

UK INTERMEDIATE MATHEMATICAL CHALLENGE

THURSDAY 3RD FEBRUARY 2000

Organised by the **United Kingdom Mathematics Trust from the School of Mathematics, University of Leeds**

RULES AND GUIDELINES (to be read before starting)

1. Do not open the paper until the Invigilator tells you to do so.
2. Time allowed: **1 hour**.
   No answers, or personal details, may be entered after the allowed hour is over.
3. The use of rough paper is allowed; **calculators** and measuring instruments are **forbidden**.
4. Candidates in England and Wales must be in School Year 11 or below.
   Candidates in Scotland must be in S4 or below.
   Candidates in Northern Ireland must be in School Year 12 or below.
5. **Use B or HB pencil only**. Mark *at most one* of the options A, B, C, D, E on the Answer Sheet for each question. Do not mark more than one option.
6. *Do not expect to finish the whole paper in 1 hour.* Concentrate first on Questions 1-15. When you have checked your answers to these, have a go at some of the later questions.
7. Five marks are awarded for each correct answer to Questions 1-15.
   Six marks are awarded for each correct answer to Questions 16-25.
   Each incorrect answer to Questions 16-20 loses 1 mark.
   Each incorrect answer to Questions 21-25 loses 2 marks.
8. Your Answer Sheet will be read only by a *dumb machine*. **Do not write or doodle on the sheet except to mark your chosen options.** The machine 'sees' all black pencil markings even if they are in the wrong places. If you mark the sheet in the wrong place, or leave bits of rubber stuck to the page, the machine will 'see' a mark and interpret this mark in its own way.
9. The questions on this paper challenge you to **think**, not to guess. You get more marks, and more satisfaction, by doing one question carefully than by guessing lots of answers. The UK IMC is about solving interesting problems, not about lucky guessing.

1. 567 is multiplied by 3489. What is the units digit of the answer?

A 1 B 3 C 5 D 7 E 9

2. An ice cream stall sells vanilla, strawberry and chocolate ice creams. The pie chart illustrates the sales of ice cream for last Saturday. The number of vanilla and the number of chocolate ice creams sold were the same. The stall sold 60 strawberry ice creams. How many chocolate ice creams were sold?

A 90 B 99 C 100 D 120 E 135

3. Which is the largest of these fractions?

A $\frac{7}{15}$ B $\frac{3}{7}$ C $\frac{11}{23}$ D $\frac{4}{9}$ E $\frac{6}{11}$

4. In Worcestershire, Wyre Piddle is 12km south of the village of North Piddle and Abbotts Morton is 12km east of North Piddle. What is the direction of Abbotts Morton from Wyre Piddle?

A South East B South West C North East D North West E West

5. In a magic square, each row, each column and both main diagonals have the same total. In the partially completed magic square shown, what number should replace x?

18			
13	15		
	10	11	17
	x	16	14

A 15 B 18 C 21 D 24 E 27

6. Granny has been having a smashing time. Yesterday she had 12 cups and 10 matching saucers, but this morning she dropped a tray holding one third of the cups and half the saucers, breaking all of those on the tray. How many of her cups are now without saucers?

A 1 B 3 C 4 D 5 E 6

7. Given that x and y are positive whole numbers and $x^2 + 2 = y^3$, which of the following is a possible value of x?

A 2 B 3 C 4 D 5 E 6

8. In the triangle ABC, $AD = BD = CD$. What is the size of angle BAC?

A 60° B 75° C 90° D 120°

E more information is needed

9. Leap years normally occur every four years. However, years at the turn of a century are leap years only if they are multiples of 400. Therefore this year, 2000, is a leap year, but the year 1900 was not a leap year. How many leap years will there be between 2001 and 3001?

A 240 B 242 C 248 D 249 E 250

10. The average (mean) weight of five giant dates was 50g. Kate ate one and the average (mean) weight of the four remaining dates was 40g. What was the weight of the date that Kate ate?

A 10 g B 50 g C 60 g D 90 g E more information is needed

11. My bargain settee cost me £240 in a sale offering 25% reductions on all items. How much did I save?

A £25 B £40 C £60 D £80 E £100

12. Timmy, Tammy and Tommy all have tummy ache! They all set off separately to visit their doctor, leaving their homes at exactly the same time. Timmy cycles the 8 km to the surgery at an average speed of 20 km/hr; Tammy walks the 1.2 km to the surgery at an average speed of 4 km/hr and Tommy drives the 16.5 km to the surgery at an average speed of 45 km/hr. In what order do they arrive at the surgery?

A Tommy, Timmy, Tammy B Timmy, Tommy, Tammy C Timmy, Tammy, Tommy

D Tammy, Timmy, Tommy E Tammy, Tommy, Timmy

13. The diagram shows two rectangles which enclose five regions. What is the largest number of regions which can be enclosed by any two rectangles drawn on a sheet of paper?

A 10 B 9 C 8 D 7 E 6

1
2 3 4
5

14. The ratio $a : b = 2 : 3$ and the ratio $a : c = 3 : 4$. What is the ratio $b : c$?

A 1 : 8 B 1 : 2 C 8 : 9 D 9 : 8 E 2 : 1

15. In how many whole numbers between 100 and 999 is the middle digit equal to the sum of the other two digits?

A 28 B 36 C 45 D 50 E 55

---

16. The pattern 123451234512345... is continued to form a 2000-digit number. What is the sum of all 2000 digits?

A 6000 B 7500 C 30 000 D 60 000 E 75 000

17. Baldrick can afford to buy either 6 parsnips and 7 turnips or else 8 parsnips and 4 turnips. Both options leave him with no change whatsoever. If, however, he bought only his beloved turnips, how many could he afford?

A 11 B 12 C 13 D 16 E 25

18. The number $3^4 \times 4^5 \times 5^6$ is written out in full. How many zeros are there at the end of the number?

A none B 4 C 5 D 6 E more than 6

19. The product of Mary's age in years on her last birthday and her age now in complete months is 1800. How old was Mary on her last birthday?

A 9 B 10 C 12 D 15 E 18

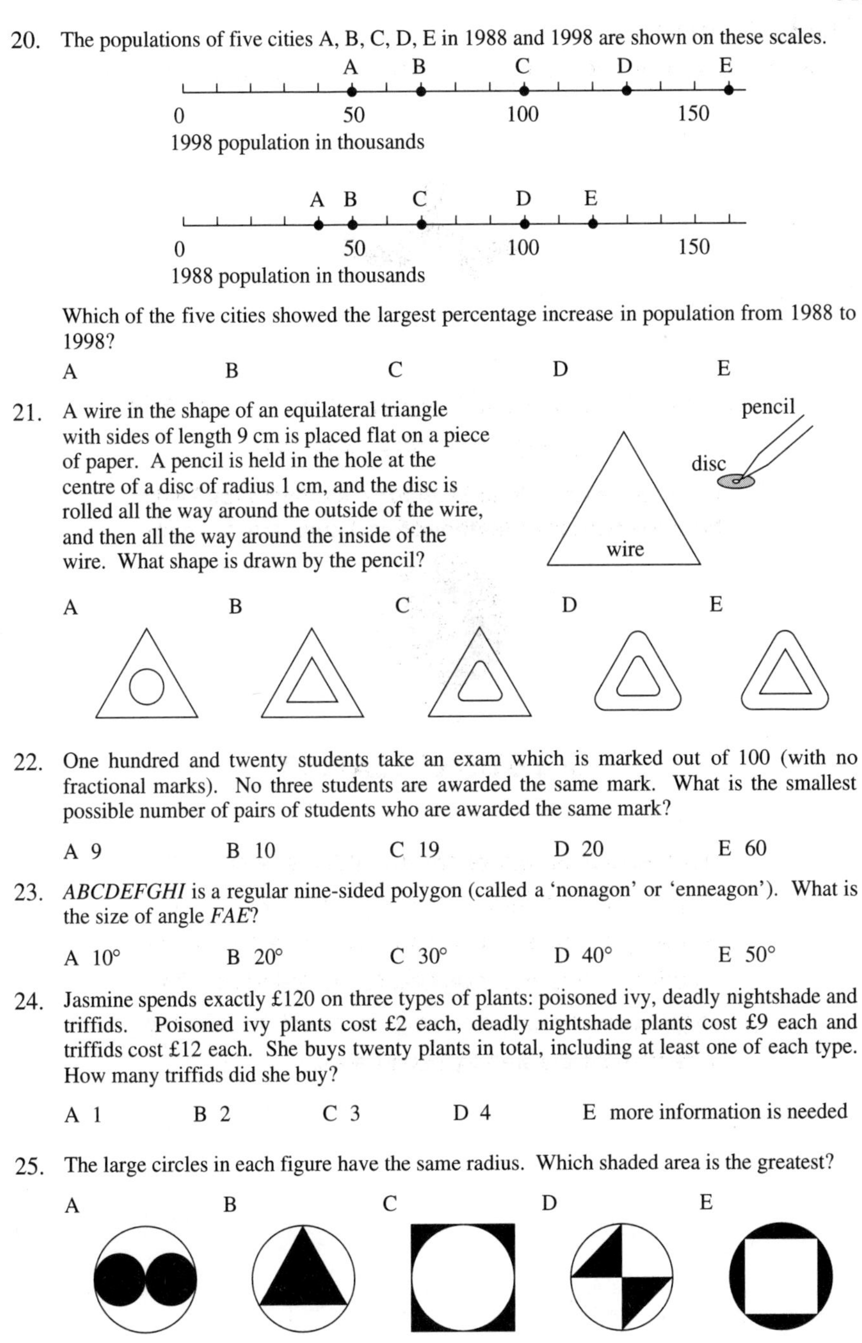

20. The populations of five cities A, B, C, D, E in 1988 and 1998 are shown on these scales.

Which of the five cities showed the largest percentage increase in population from 1988 to 1998?

A   B   C   D   E

21. A wire in the shape of an equilateral triangle with sides of length 9 cm is placed flat on a piece of paper. A pencil is held in the hole at the centre of a disc of radius 1 cm, and the disc is rolled all the way around the outside of the wire, and then all the way around the inside of the wire. What shape is drawn by the pencil?

A   B   C   D   E

22. One hundred and twenty students take an exam which is marked out of 100 (with no fractional marks). No three students are awarded the same mark. What is the smallest possible number of pairs of students who are awarded the same mark?

A 9   B 10   C 19   D 20   E 60

23. *ABCDEFGHI* is a regular nine-sided polygon (called a 'nonagon' or 'enneagon'). What is the size of angle *FAE*?

A 10°   B 20°   C 30°   D 40°   E 50°

24. Jasmine spends exactly £120 on three types of plants: poisoned ivy, deadly nightshade and triffids. Poisoned ivy plants cost £2 each, deadly nightshade plants cost £9 each and triffids cost £12 each. She buys twenty plants in total, including at least one of each type. How many triffids did she buy?

A 1   B 2   C 3   D 4   E more information is needed

25. The large circles in each figure have the same radius. Which shaded area is the greatest?

A   B   C   D   E

The IMC solutions

As with the Junior Challenge, a solutions leaflet was sent out.

UK INTERMEDIATE MATHEMATICAL CHALLENGE

THURSDAY 3RD FEBRUARY 2000

Organised by the **United Kingdom Mathematics Trust from the School of Mathematics, University of Leeds**

SOLUTIONS LEAFLET

This solutions leaflet for the IMC is sent in the hope that it might provide all concerned with some alternative solutions to the ones they have obtained. It is not intended to be definitive. The organisers would be very pleased to receive alternatives created by candidates.

1. **B** Take the units digits of any two numbers and multiply them together. The units digit of the answer is also the units digit of the product of the original two numbers. As $7 \times 9 = 63$, the units digit of 567×3489 must also be 3.

2. **A** The angle occupied by the 'chocolate' sector is $\frac{1}{2}(360° - 90°) = 135°$. This is $1\frac{1}{2}$ times bigger than the 'strawberry' sector and hence the number of chocolate ice creams sold is $1\frac{1}{2} \times 60 = 90$.

3. **E** $\frac{6}{11}$ is the only one of these fractions which is greater than $\frac{1}{2}$.

4. **C**

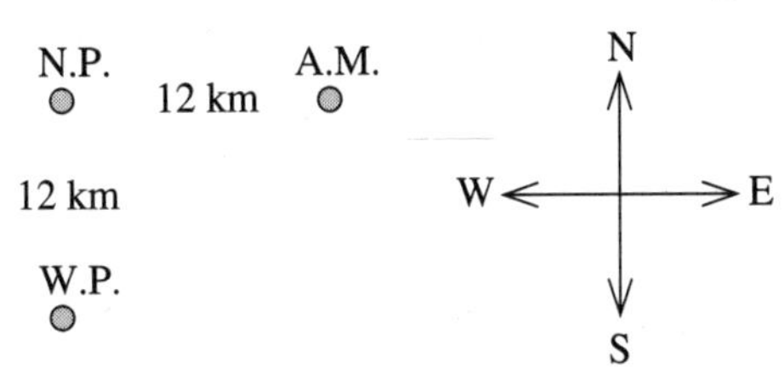

5. **C** The numbers along the leading diagonal total 58 and this is therefore the sum of each row and column. We can now calculate that the number to the left of the '10' must be 20 and the number below that is 7.
Hence $x = 58 - (16 + 14 + 7) = 21$.

(A more difficult task is to calculate the value of the number in the top right-hand corner of the magic square. Can you do this?)

6. **B** Granny dropped 4 cups and 5 saucers, leaving her with 8 cups and 5 saucers. Therefore 3 cups did not have matching saucers.

7. **D** $5^2 + 2 = 3^3$

(This is in fact the only solution of this equation for which x and y are positive whole numbers. Another way of looking at this is to say that 26 is the only whole number which is 'sandwiched' between a perfect square and a perfect cube. This was first proved by the French mathematician, Pierre de Fermat, in the 17th century.)

8. **C** A, B and C are all equidistant from D and therefore lie on a circle whose centre is D. BC is a diameter of the circle and $\angle BAC$ is therefore the angle subtended by a diameter at a point on the circumference (the angle in a semicircle).
(Alternatively: let $\angle ACD = x$ and show that $\angle DAC = x$, $\angle ADB = 2x$ and $\angle DAB = \frac{1}{2}(180° - 2x) = 90° - x$. Hence $\angle BAC = x + 90° - x = 90°$.)

9. **B** The numbers of multiples of 4 between 2001 and 3001 is 250. However, the following years will not be leap years: 2100, 2200, 2300, 2500, 2600, 2700, 2900, 3000. This leaves 242 leap years.

10. **D** The total weight of the original five dates was 250g and the total weight of the four remaining dates was 160g.

11. D The sale price is 75% of the original price. Therefore the amount I saved, 25% of the original price, is one third of £240, i.e. £80.

12. E Timmy takes 24 minutes (8/20 of 1 hour) to reach the surgery; Tammy takes 18 minutes (12/40 of 1 hour) and Tommy takes 22 minutes (33/90 of 1 hour). The order, therefore, is Tammy, Tommy, Timmy.

13. B

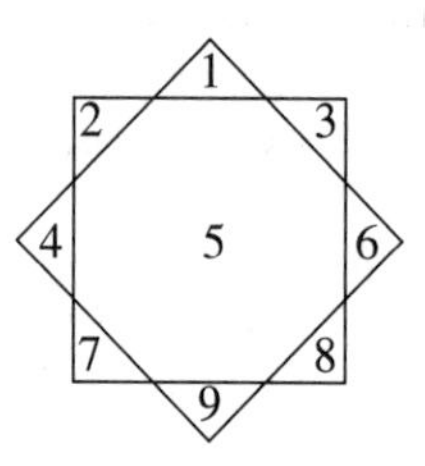

14. D $\frac{b}{c} = \frac{a}{c} \times \frac{b}{a} = \frac{3}{4} \times \frac{3}{2} = \frac{9}{8}$.

15. C There are 9 such numbers whose first digit is 1 : 110, 121, 132, …, 187, 198. Similarly there are 8 such numbers, beginning with 220 and ending with 297, whose first digit is 2; 7 such numbers, beginning with 330 and ending with 396, whose first digit is 3 and so on. Lastly there is only 1 such number whose first digit is 9: 990.

The answer, therefore, is $9 + 8 + 7 + 6 + 5 + 4 + 3 + 2 + 1 = 45$.

16. A The number may be divided up into 400 blocks of '12345'. The sum of the digits in each block is 15 and hence the sum of all 2000 digits is $400 \times 15 = 6000$. Alternatively, the mean of the digits which make up the number is 3 and therefore the sum of the digits is $2000 \times 3 = 6000$.

17. D Buying two more parsnips and three fewer turnips does not change the total cost and hence two parsnips cost the same as three turnips. Instead of six parsnips, therefore, Baldrick could have bought nine turnips and, together with seven turnips, this makes a total of sixteen turnips. Alternatively, he could have bought twelve turnips instead of the eight parsnips and, together with four turnips, this makes sixteen turnips.

18. D We need to write $3^4 \times 4^5 \times 5^6$ in the form $a \times 10^n$ where a is not a multiple of 10.
$3^4 \times 4^5 \times 5^6 = 3^4 \times 2^{10} \times 5^6 = 3^4 \times 2^4 \times 2^6 \times 5^6 = 3^4 \times 2^4 \times 10^6$.
Hence the number ends in six zeros.

19. C We need to express 1800 as the product of two factors, one of which (her age in months) is between twelve and thirteen times the other (her age in complete years). These are 150 and 12 respectively. Mary is 150 months old i.e. she was twelve on her last birthday and she is now 12 years 6 months old.

20. C The percentage increases are A 25%; B 40%; C $42\frac{6}{7}$%; D 30%; E $33\frac{1}{3}$%

21. E On the outside of the wire, the pencil describes an arc of a circle as the disc rolls around each of the corners of the triangle, but this does not happen when the disc moves around the inside of the wire.

22. C The smallest possible number of pairs of students with the same mark will occur when every possible mark from 0 to 100 is awarded to at least one student. This accounts for 101 students and therefore the remaining 19 students must all be awarded the same mark as exactly one of their colleagues. The 120 students are made up of 19 pairs of students who are awarded the same mark and 82 students who are all awarded a different mark from everyone else.

23. B The interior angle of a regular nine-sided polygon $= 180° - (360° \div 9)$ $= 140°$. Consider the pentagon *ABCDE*: $\angle EAB = \frac{1}{2}(540° - 3 \times 140°) = 60°$. Similarly, $\angle FAI = 60°$ and hence $\angle FAE = 140° - (60° + 60°) = 20°$.

24. A Let the number of ivy, nightshade and triffid plants be i, n and t respectively. Then:
$2i + 9n + 12t = 120$ and $i + n + t = 20$, where $i > 0; n > 0; t > 0$. Multiplying the second equation by 2 and subtracting the new equation from the first:

$$7n + 10t = 80$$

$$\text{Thus } 7n = 10(8 - t)$$

Therefore n is a multiple of 10 and since $1 \leqslant n < 20, n = 10$.
Hence $8 - t = 7$ and therefore $t = 1$.

25. A Let the large circles have radius R.
Area $A = 2 \times \pi \times (\frac{1}{2}R)^2 = \frac{1}{2}\pi R^2 \approx 1.6R^2$.
Area $B = 3 \times \frac{1}{2} \times R \times R \times \sin 120° = \frac{3}{4}\sqrt{3}R^2 \approx 1.3R^2$.
Area $C = (2R)^2 - \pi R^2 = (4 - \pi)R^2 \approx 0.9R^2$.
Area $D = 2 \times \frac{1}{2} \times R^2 = R^2$.
Area $E = \pi R^2 - (\sqrt{2}R)^2 = (\pi - 2)R^2 \approx 1.1R^2$.

The answers

The table below shows the proportion of pupils' choices. The correct answer is shown in bold.

Q	A	B	C	D	E	Blank
1	4	**54**	8	27	5	2
2	**68**	1	4	11	14	1
3	3	22	9	6	**57**	3
4	11	11	**70**	6	1	1
5	11	17	**58**	5	3	6
6	10	**80**	5	3	1	1
7	18	6	12	**54**	5	4
8	11	15	**10**	4	54	5
9	9	**17**	12	18	37	6
10	41	8	7	**26**	15	3
11	2	4	40	**49**	1	3
12	19	15	9	16	**36**	4
13	4	**42**	12	21	15	4
14	4	26	18	**32**	10	9
15	11	20	**36**	11	10	11
16	**51**	3	16	6	2	23
17	10	8	6	**27**	3	47
18	24	10	4	**6**	12	43
19	3	3	**14**	17	6	56
20	3	5	**16**	2	42	31
21	3	7	11	27	**20**	31
22	4	10	**5**	15	7	58
23	2	**7**	5	21	2	62
24	**6**	3	4	6	20	60
25	**14**	18	4	2	2	60

IMC 2000: Some comments on the pupils' choices of answers

The average score, 39.3, was a little lower than last year, and there were several surprises for the Problems Group. The intention is to start with an very easy question, and it was therefore disappointing that only 54% of the entrants got Question 1 correct. We were puzzled to see that 27% of the entrants chose the answer D. Did they not understand what is meant by 'units digit'? It has been suggested that D was popular because the answer to the sum is a 7 digit number.

It is the questions where one particular wrong answer turns out to be more popular than the correct answer which give us most cause for concern. 'More information is needed' is always a popular answer, and this was shown this year by the responses to Question 8, where over half the entrants chose this easy way out.

In both Questions 9 and 18 careful thought is needed to get the right answers, but in each case the most popular wrong answer was the one which should have been the easiest to rule out!

We emphasize that 'the questions challenge you to think, not to guess'. However, many entrants do not read the questions carefully enough. Instead they often seize on some of the numbers given in the question, do a calculation which leads to one of the suggested answers, and then choose this answer without stopping to think whether the calculation they did was relevant to the question they were asked.

For example, in Question 10, 50g - 40g = 10g, and 41% of the entrants chose this answer. In question 11, 25% of £240 is £60, and 40% chose this answer.

In contrast, there were a few questions which were done better than we expected. Question 6 is a notable example. In needs careful thought to disentangle the correct answer from the information you are given. So it was pleasing that 80% of the entrants chose the right answer. Perhaps the subject matter was closer to their experience than that of the other questions.

The profile of marks obtained is shown below.

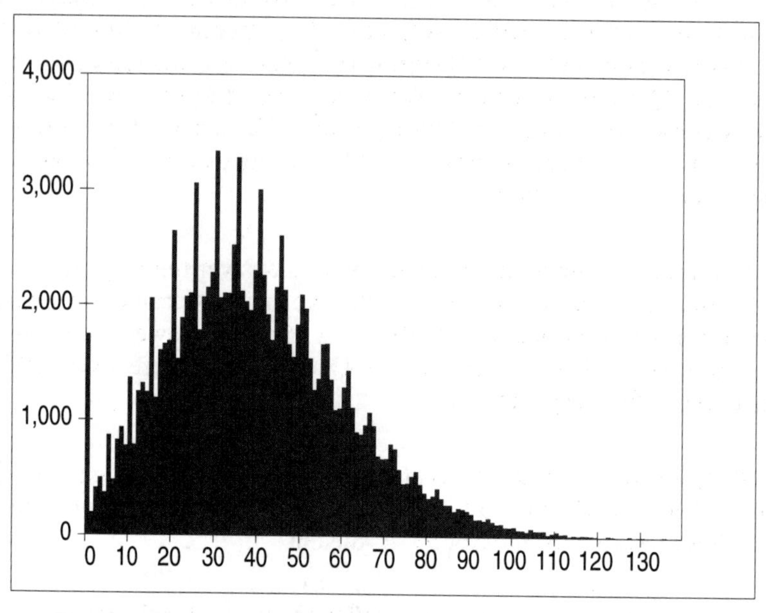

Barchart showing the actual frequencies in the 2000 IMC

On the basis of the standard proportions used by the UKMT, the cut-off marks were set at

GOLD – 73 or over SILVER – 55 to 72 BRONZE – 40 to 54

The certificates were virtually identical to those used for the JMC.

The cut-off scores for the follow-up competitions were

IIIMC (11)	Pupils in year 11 scoring 97 or more.
IIIMC (10)	Pupils in year 10 scoring 95 or more.
Kangaroo	Pupils in year 9 scoring 65 or more.

The European Kangaroo – 'Kangourou sans Frontières'

The 2000 European Kangaroo took place on Thursday 16th March. It was also held in over 20 countries across Europe. It is a multiple choice paper with 25 questions. (It is only one of a number of similar competitions taking place across Europe.)

EUROPEAN 'KANGAROO' MATHEMATICAL CHALLENGE

THURSDAY 16TH MARCH 2000

Organised by the United Kingdom Mathematics Trust and the Association Kangourou des Mathématiques, Paris

This paper is being taken by students in twenty European countries.

RULES AND GUIDELINES (to be read before starting):

1. Do not open the paper until the Invigilator tells you to do so.
2. Time allowed: **1 hour**
   No answers, or personal details, may be entered after the allowed hour is over.
3. The use of **calculators**, rulers and measuring instruments is **forbidden**.
4. Candidates in England and Wales must be in School Year 9 or below.
   Candidates in Scotland must be in S2 or below.
   Candidates in Northern Ireland must be in School Year 10 or below.
5. **Use B or HB pencil only**. For each question mark *at most one* of the options A, B, C, D, E on the Answer Sheet. Do not mark more than one option.
6. Five marks will be awarded for each correct answer to Questions 1 - 15.
   Six marks will be awarded for each correct answer to Questions 16 - 25.
7. *Do not expect to finish the whole paper in 1 hour.* Concentrate first on Questions 1-15. When you have checked your answers to these, have a go at some of the later questions.
8. The questions on this paper challenge you **to think**, not to guess. You get more marks and more satisfaction, by doing one question carefully than by guessing lots of answers.

Enquiries about the European Kangaroo should be sent to: Maths Challenges Office, School of Mathematics, University of Leeds, Leeds, LS2 9JT.
(Tel. 0113 233 2339)

1. The diagram shows a clock seen in a mirror. What time is on the clock?

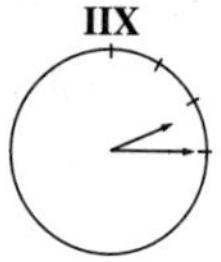

   A 15:15 B 10:15 C 10:45 D 8:45 E 9:45

2. 80% of a black-and-white photograph of Kanga is black and 20% is white. Kanga's photograph is enlarged to three times its original size. What percentage of the enlarged photograph is white?

   A 20% B 30% C 40% D 60% E 80%

3. What is the maximum number of shapes like this 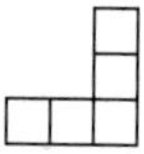 that can be placed within the square on the right without overlapping?

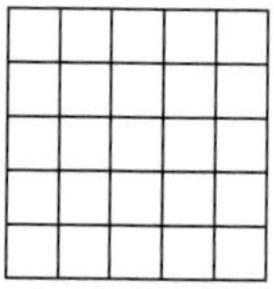

   A 2 B 3 C 4 D 5 E 6

4. If all the diagonals of a regular hexagon are drawn. how many points of intersection are there, not counting the corners of the original hexagon?

   A 6 B 7 C 12 D 13 E 15

5. On a thin strip of paper marks are drawn dividing the strip into 4 equal lengths. Marks are also drawn dividing the strip into 3 equal lengths. After that, the strip is cut in accordance with all the marks. How many different lengths will the pieces have?

   A 2 B 3 C 4 D 5 E 6

6. The sum of seven consecutive odd numbers is 119. What is the smallest of these numbers?

   A 11 B 13 C 15 D 17 E 19

7. $AD = DC$, $AB = AC$, $\angle ABC = 75°$ and $\angle ADC = 50°$. What is the size of $\angle BAD$?

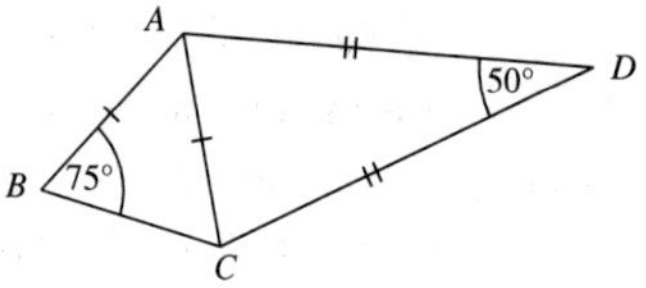

   A 30° B 85° C 95°
   D 125° E 140°

8. The most experienced trainer at the circus needs 40 minutes to wash an elephant. His young son completes the same task in 2 hours. How long will it take the trainer with his son to wash three elephants, if they work together?

   A 30 minutes B 45 minutes C 60 minutes D 90 minutes E 100 minutes

9. What size is the area shown shaded in the diagram?

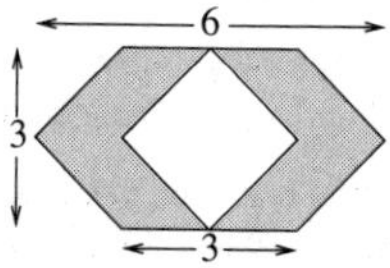

   A 9 B $3\sqrt{2}$ C 18
   D 12 E $6\sqrt{3} - 3\sqrt{2}$

10. Five people P, Q, R, S and T shook hands with each other. P shook hands once. Q also shook hands once, and each of R, S and T shook hands twice. It is known that P shook hands with T. Which handshake certainly did not occur?

    A T with S B T with R C Q with R D Q with T E Q with S

11. The area of a sector of a circle is 15% of the area of the whole circle. What is the angle of the sector at the centre of the circle?

A 15° B 36° C 54° D 90° E 150°

12. 800 grosses have the same value as 100 ducats. 100 grosses have the same value as 250 tollars. How many ducats have the same value as 100 tollars?

A 2 B 5 C 10 D 25 E 50

13. Roo's mother bought a full box of sugar cubes, the box being in the shape of a cuboid. Roo ate the top layer, which contained 77 cubes. Later, he ate the layer on the right-hand side, which had 55 cubes. Finally, he ate the layer at the front. How many cubes were left in the box?

A 203 B 256 C 295 D 300 E 350

14. At a dancing competition each of the judges grades each competitor with a whole number. Kanga takes part and the mean of all her grades is 5·625. What is the smallest number of judges for which this is possible?

A 2 B 6 C 8 D 10 E 12

15. In a certain National Park in Australia where kangaroos live, we know that:

if the sun is shining, then the temperature is not below 25°;

if the temperature exceeds 26°, then the sun is shining.

Which of the following statements necessarily follows?

A The night-time temperature is below 25°

B The daytime temperature is above 24°

C The night-time temperature cannot be 27°

D The daytime temperature cannot be 24°

E If the temperature is 25° then the sun is shining.

16. Each of the nets below has been coloured with two colours. Which net can be used to make a cube so that any two regions which have an edge in common are of the same colour?

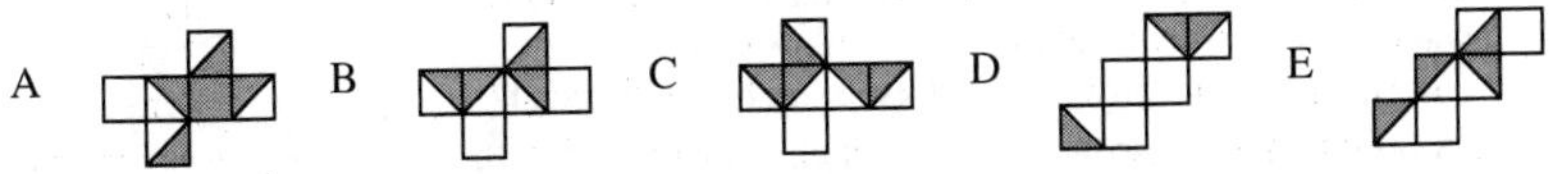

A B C D E

17. In three years from now Steve will be three times as old as he was three years ago. In four years from now Steve will be ... as old as he was four years ago. What are the missing words?

A two times B three times C four times D five times E six times

18. Points P, Q, R and S divide the sides of a rectangle $ABCD$ in the ratio $1:2$ as shown in the figure. What fraction of the area of the rectangle is the area of parallelogram $PQRS$?

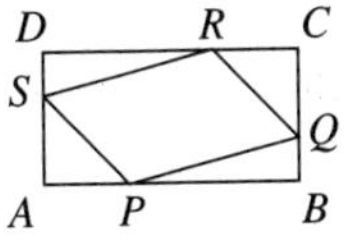

D R C S Q A P B

A $\frac{2}{5}$ B $\frac{3}{5}$ C $\frac{4}{9}$ D $\frac{5}{9}$ E $\frac{2}{3}$

19. Roo was given a box containing 2000 sweets of 5 colours. 387 of them were white, 396 yellow, 402 red, 407 green and 408 brown. Roo decided to eat the sweets as follows: without looking at them he randomly took 3 sweets from the box. If they were all the same colour he ate them, otherwise he put them back in the box. He continued in this way all day long. In the evening only 2 sweets of the same colour remained in the box. What colour were they?

    A white B yellow C red D green E brown

20. Kanga has a large number of building blocks which are all cuboids measuring 2 cm by 6 cm by 1 cm. She wants to use some of the blocks to make a cube. What is the smallest number of blocks she requires?

    A 6 B 12 C 18 D 36 E 144

21. Excluding 1 and 24 itself, the positive whole numbers which divide 24 exactly are 2, 3, 4, 6, 8 and 12. These six numbers are called the *proper divisors* of 24. Suppose that you wanted to write down in increasing order all the whole numbers n greater than 1 for which n is equal to the product of its proper divisors. What would be the sixth number that you wrote down?

    A 14 B 15 C 21 D 22 E 25

22. A magic rectangular piece of "Shagreen" leather reduces by $\frac{1}{2}$ of its length and by $\frac{1}{3}$ of its width each time it fulfils a wish of its owner. After 3 wishes it was of area 4 cm^2, and its initial width was 9 cm. What was its initial length?

    A 12 cm B 36 cm C 4 cm D 18 cm E impossible to determine

23. Roo had 6 wooden rods, and entertained himself by using all of them to form equilateral triangles, so that the rods only touched at their ends. One day he lost one of the rods. Since he really enjoyed this game, he asked his father to make a new rod for him. How many possibilities are there for the length of the new rod, if the lengths of the other rods are 25, 29, 33, 37 and 41?

    A 1 B 2 C 3 D 4 E 5

24. Nine different dominoes form a shape, partly covered by a piece of paper. The dominoes meet one another in such a way that 1 dot is adjacent to 1 dot, 2 dots are adjacent to 2 dots, and so on. How many dots are in the shaded cell? [A domino has two cells. each of which is marked with 0, 1, 2, 3, 4, 5 or 6 dots.]

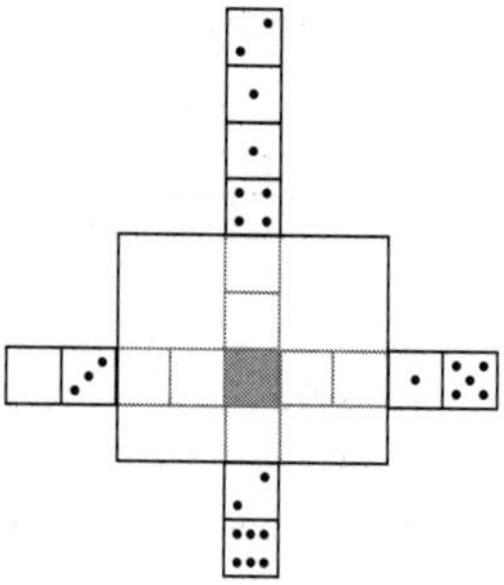

    A 2 B 3 C 4
    D another answer E impossible to determine

25. What is the last digit in the finite decimal representation of the number $\frac{1}{5^{2000}}$?

    A 2 B 4 C 6 D 8 E 5

European 'Kangaroo' Mathematical Challenge
March 2000 Solutions

1. E When the clock has been reflected back to its original position, the hour hand is then between 9 and 10 and the minute hand points at the 9 exactly.
The time on the clock is therefore 9:45.

2. A Enlarging the photograph does not change the relationship between black and white and so the percentage remains 20%.

3. C As shown, four of the 'L' shape tiles can be fitted into the square. But, how do we know that four is the best possible solution?

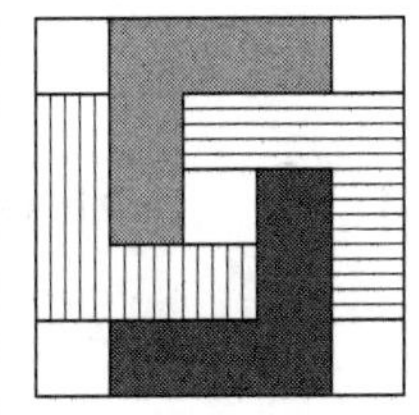

Each tile has a vertical arm and a horizontal arm – each of which must intersect the middle row or middle column respectively. Suppose you have four tiles in somehow. You cannot have one of the tiles having an arm running along the middle row (or column) – since that would use three of its squares which would not leave enough room for the other three tiles to cross it. So the middle square cannot be used in any solution having four or more tiles. This leaves only four squares and each tile needs five – so a fifth tile canot be fitted.

4. D

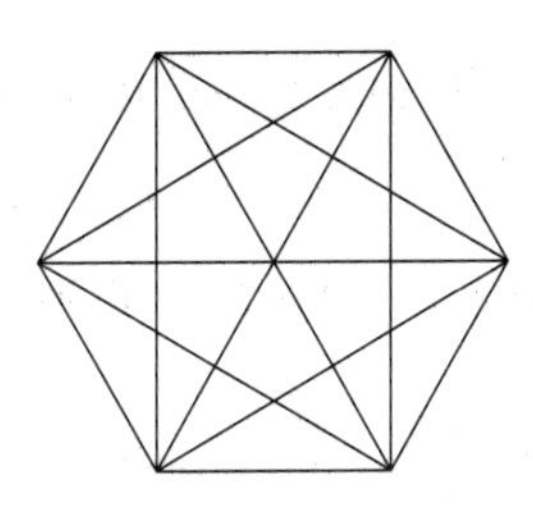

5. B

3	3	3	3

4	4	4

3	1	2	2	1	3

Since we are to divide the strip by four and by three, we can say the strip originally had 12 'parts' – giving four parts of length 3 and three parts of length 4. When the second set of cuts is made, the strip is cut into lengths 1, 2 and 3 as shown above.

6. **A** Every even number can be expressed as $2n$ and an odd number is always one more than an even number and so can be written as $2n + 1$. Let this be the middle number of the seven consecutive odd numbers. Then all seven numbers are

$$2n - 5,\ 2n - 3,\ 2n - 1,\ 2n + 1,\ 2n + 3,\ 2n + 5,\ 2n + 7$$

which all add to $14n + 7$. Therefore $14n + 7 = 119$, $14n = 112$, $n = 8$ and the smallest number is 11.

7. **C**

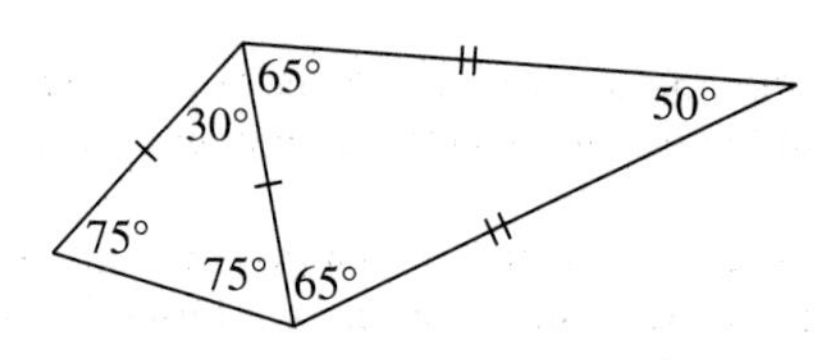

8. **D** In 2 hours = 120 minutes, the father and son together wash 4 elephants. So they wash 3 elephants in $\frac{3}{4} \times 120 = 90$ minutes.

9. **A** The shaded area can be split into 8 identical isosceles right-angled triangles each with base and height 1·5 units.

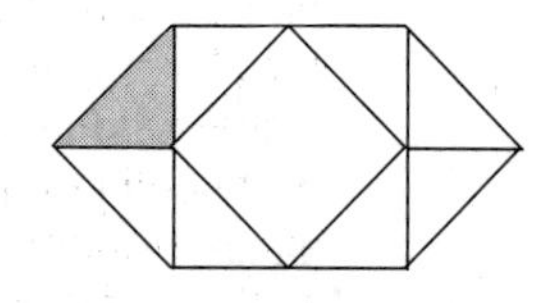

The shaded area = $8 \times \frac{1}{2} \times \frac{3}{2} \times \frac{3}{2} = 9$.

10. **D** If Q shook hands with T, then R and S would have to shake hands with each other twice.

11. **C** The angle at the centre is 15% of 360° $= \frac{15}{100} \times 360°$.

12. **B** 100 tollars is $\frac{100}{250} \times 100$ grosses $= 40$ grosses.

40 grosses is $\frac{40}{800} \times 100$ ducats $= 5$ ducats.

13. **D** The original top layer was 7 cubes wide and 11 cubes broad. After the top layer was removed, the side was 5 cubes high and 11 cubes broad and so was originally 6 cubes high and 11 cubes broad. After the top and the right hand side had been removed, the front was 6 cubes wide and 5 cubes tall and so 30 cubes were eaten. The number of cubes left was $(11 \times 7 \times 6) - (77 + 55 + 30)$.

Alternatively, notice that the final pile is 6 cubes wide, 10 cubes long and 5 cubes high and $6 \times 10 \times 5 = 300$.

14. C Since $0{\cdot}625 = \frac{5}{8}$, the smallest number that can multiply 5.625 to make a whole number is 8.

15. C If the temperature exceeds 26° then the sun is shining. So when the temperature is 27° the sun is shining which is not possible during the night.

16. D

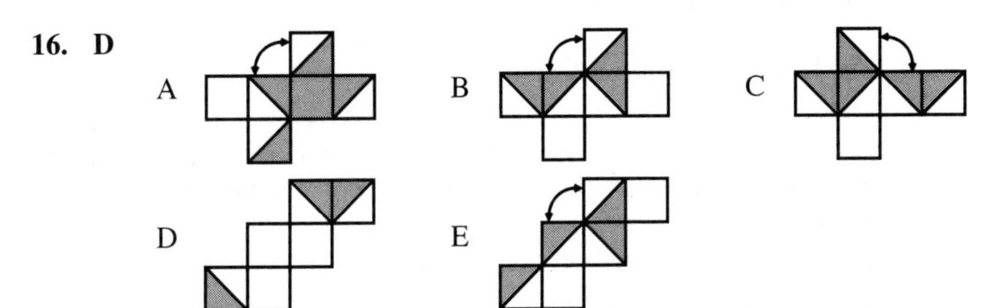

The arrows indicate edges at which the coloured faces do not match as required.

17. D Let Steve's present age be x.
Then $x + 3 = 3(x - 3)$ and Steve's present age is 6.
Four years ago he was 2 and he will be 10 in four years time.

18. D

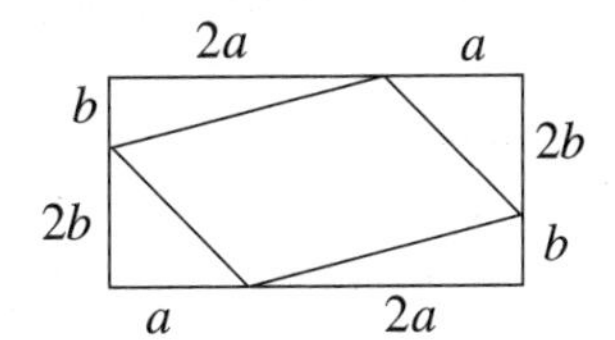

Area of rectangle = $3a \times 3b$
Area of each triangle = ab
Area of parallelogram
$= 9ab - 4ab = 5ab.$

19. D Roo only ate three sweets at a time. If there were two sweets left, there must have been two plus a multiple of three of that colour at the start.

$387 = 3 \times 129$ $396 = 3 \times 132$ $402 = 3 \times 134$
$408 = 3 \times 136$

and $407 = 2 + 3 \times 135$.
Or, you can notice that 407 is the only number offered which leaves a reminder of 2 when it is divided by 3.

20. C The side of the final cube must be at least 6 cm long. She can make a 6 cm cube by arranging 6 sides of 1 cm together and then placing 3 three layers of these (which are 2 cm high) on top of one another.

21. C The required numbers are 6, 8, 10, 14, 15, 21, … .

6 = 2 × 3	**8** = 2 × 4	**10** = 2 × 5
14 = 2 × 7	**15** = 3 × 5	**21** = 3 × 7

22. A After three washes, the width became $\frac{2}{3} \times \frac{2}{3} \times \frac{2}{3} \times 9 = \frac{8}{3}$ cm.

The length was then $4 \div \frac{8}{3} = \frac{3}{2}$ cm.

and the original length was $2 \times 2 \times 2 \times \frac{3}{2} = 12$ cm.

23. C You must use exactly two rods to make each side of the triangle (because 41 is too small to make a single side and the sum of three of the numbers is too big).
Consider pairs that make the same total

25 + 41 = 29 + 37 = 33 + **33** = 66 25 + 37 = 29 + 33 = 41 + **21** = 62

25 + 33 = 58 and 25 + 29 = 54 are too small to find any combinations of other numbers to suit. However, 25 + **45** = 29 + 41 = 33 + 37 = 70.

The only possible lengths are 45, 33 and 21.

24. B Consider all the remaining dominoes with one dot – to place at the right hand side ⚀⚀, ⚀⚂, ⚀⚅, ⚀□ which lead to the positions

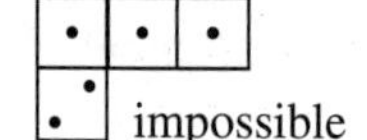
impossible

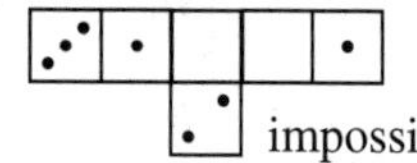
impossible

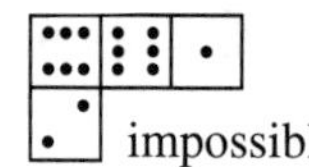
impossible

or: which is the only possible solution.

25. C $\frac{1}{5} = 0{\cdot}2$ $\frac{1}{5^2} = 0{\cdot}2^2 = 0{\cdot}04$ $\frac{1}{5^3} = 0{\cdot}2^3 = 0{\cdot}008$ $\frac{1}{5^4} = 0{\cdot}2^4 = 0{\cdot}0016$

2^1	2^2	2^3	2^4	2^5	2^6	2^7	2^8	2^9	2^{10}	2^{11}	2^{12}
2	4	8	16	32	64	128	256	512	1024	2048	4096

Notice that the final digit corresponds with the final digit in a power of 2 and that the powers of 2 end in a sequence

2, 4, 8, 6 2, 4, 8, 6 2, 4, 8, 6

$5^{2000} = 5^{4 \times 500}$ and has final digit the same as every power of 2^4.

Kangaroo awards and high scorers

As in the JMO, everyone is awarded a certificate with about a quarter obtaining a Certificate of Distinction and the rest a Certificate of Participation. In addition, the top 100 received a copy of the French magazine *Les malices du kangourou.*

The high scorers were

135	Stephen Legg	Toot Hill CS
130	Paul Jefferys	Berkhamsted Collegiate
125	Alex Smith	KE VI Five Ways School
124	Anton Baker	Westminster School
	Duncan Cameron	Framingham Earl High
	Alex Davies	Kings School Bruton
	Gary Kendall	Hinchingbrooke School
	Liam Taylor	Bristol Grammar
	Bingyuan Yang	Loughborough Grammar
119	Giles Coope	Stockport Grammar School
	Oliver Hallam	The Perse School
	Martin Orr	Methodist College
	Robert Sparkes	Hobart HS
118	Richard Brewis	Bedford School
	Peter Coulthard	Portsmouth Grammar
	Michael Donaghy	St Albans HS Ipswich
	Katie Fahey	Orton Longueville CS
	Michael Gathercol	West Somerset Minehead
	Tim Jeffries	Oundle School
	Peter Myall	King Edward VI GS
	David Tite	KES Birmingham
117	Matthew You	The Judd School

114	Ben Vail	West Exe TC
113	Aneesh Barai	Eltham College
	James Carrington	Bottisham VC
	Jonathan Cheyne	St Olaves GS
	Ben Crowne	City of London School
	Samuel Hayler	North Leamington
	Rebekah Nolan	Crofton
113	David Smy	Woolmer Hill School
	Edward Stutters	Churchers College
	Paul Thomas	Wilsons School
	Paul Totman	Wilsons School
	Timothy Woodward	Westminster School
112	Gillian Bradley	Casterton School
	Matthew Chinery	Uppingham School
	Duncan Kemp	The Perse School
	Tim Landy	King Alfreds Wantage
	Sam Lings	St Mary Redcliffe School
	Edward Nesbit	Malvern College
	Thomas Ryder	The Judd School
	Glenn Sheasby	St Richards Catholic C
	Laura Short	Oundle School
	Priya Singh	The Netherhall School
	James Staff	Devonport Boys HS

A sample of a 'Kangaroo' certificate

2000 European Kangaroo

....

of

....

received a

CERTIFICATE of PARTICIPATION

'Kangourou
sans
Frontières'

Peter Neumann

Chairman, United Kingdom Mathematics Trust

THE EUROPEAN KANGAROO

The European Kangaroo is a competition in which a million and a half pupils from 21 countries throughout Europe participate. The questions provide amusing and thought provoking situations which require the use of logic as well as numeracy skills.

In the United Kingdom, invitation to participate in the Kangaroo is an honour in recognition of previous high achievement in the Intermediate Mathematical Challenge by pupils in year 9 or below.

After Easter there were further follow-up competitions with the IIIMC 10 and 11 held on Thursday 4th May

2000 Invitational Mathematics Challenge

(Grade 10)

TIME ALLOWED: 2 hours

Rulers and compasses may be used, **but not** protractors or squared paper

Calculators are not permitted. It is expected that all calculations and answers will be expressed as exact numbers such as 4π, $2 + \sqrt{7}$, etc. Marks are awarded for completeness, clarity, and style of presentation. A correct solution poorly presented will not earn full marks.

1. Solve the system of equations:

$$x^2 + xy + y^2 = 21$$
$$x^2 - xy + y^2 = 13.$$

2. A horizontal wooden frame is made in the form of an equilateral triangle with each side 1 metre in length. Sean attaches a 6 metre rope to the frame at A and walks to point P so that P, A, and B lie on a straight line, as shown. Sean then sets off in the direction shown and continues walking, keeping the rope taut and horizontal at all times. How far does he walk after leaving P until the rope gets wound twice round the frame?

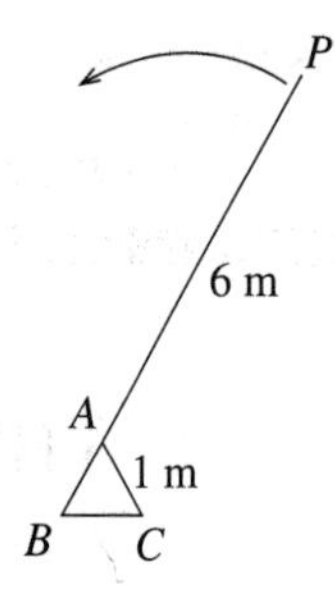

3. A line with gradient 4 is drawn through the midpoint of the line connecting $B(2, 3)$ and $C(10, 3)$. Point A is chosen on this line such that the area of triangle ABC is 32. Determine the coordinates of the two possible positions of A.

4. (a) In the given diagram, AD is perpendicular to BC, where point D is on BC. Points P and R are on AB and AC, respectively, such that making folds in the paper along PQ and RS would place both B and C on point D. Prove that a fold made along PR places vertex A on point D.

   (b) In this triangle, BC has length a, AC has length b and AB has length c. Determine the length of BQ in terms of a, b and c.

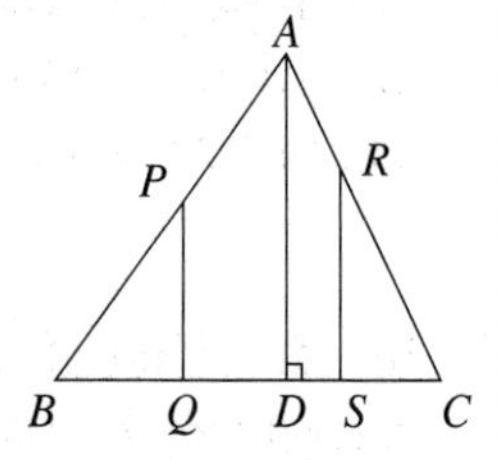

5. Determine all triples (a, b, c) of integers such that $1 \leqslant a \leqslant b \leqslant c$ and

$$\frac{1}{a+2} + \frac{1}{b+2} = \frac{1}{2} + \frac{1}{c+2}.$$

Canadian Mathematics Competition

An activity of The Centre for Education in Mathematics and Computing, Faculty of Mathematics, University of Waterloo, Waterloo, Ontario, Canada

Solutions for 2000 Invitational Mathematics Challenge (Grade 10)

---

1. The given equations are

$$x^2 + xy + y^2 = 21 \ldots (1); \qquad x^2 - xy + y^2 = 13 \ldots (2).$$

Subtract (2) from (1) to obtain $2xy = 8$, i.e. $xy = 4 \ldots (3)$.

(N.B. at this point we must ***not*** *assume that the solutions x, y are integers.)*

Add (1) to (3) to obtain

$$x^2 + 2xy + y^2 = 21 + 4, \text{ i.e. } (x + y)^2 = 25,$$

$$\text{so } x + y = 5 \ldots (4); \text{ or } x + y = -5 \ldots (5).$$

From (3): $y = \frac{4}{x}$. Substituting for y in (4) gives $x + \frac{4}{x} = 5$ i.e. $x^2 - 5x + 4 = 0$ i.e. $(x - 1)(x - 4) = 0$. Hence $x = 1$ or $x = 4$ and the corresponding values of y are 4 and 1 respectively.

Similarly, substituting for y in (5) gives $x + \frac{4}{x} = -5$ i.e. $x^2 + 5x + 4 = 0$ i.e. $(x + 1)(x + 4) = 0$. Hence $x = -1$ or $x = -4$ and the corresponding values of y are -4 and -1 respectively.

Thus there are four solution pairs (x, y), namely $(1, 4)$, $(4, 1)$, $(-1, -4)$ and $(-4, -1)$.

Note: an alternative solution for solving the simultaneous equations makes use of the formulae for the sum and product of the roots of the quadratic equation $ax^2 + bx + c = 0$: sum of roots $= -\frac{b}{a}$; product of roots $= \frac{c}{a}$.

When $x + y = 5$, use of (3) shows us that x and y are the roots of the quadratic equation $t^2 - 5t + 4 = 0$, i.e. $(t - 1)(t - 4) = 0$. Thus x and y are the numbers 1 and 4 in either order. We therefore obtain the pairs $(1, 4)$ and $(4, 1)$.

When $x + y = -5$, we see similarly that x and y are the roots of the quadratic equation $t^2 + 5t + 4 = 0$, i.e. $(t + 1)(t + 4) = 0$. This time x and y must be -1 and -4 in either order and we obtain the pairs $(-1, -4)$ and $(-4, -1)$.

---

2. Sean walks:

half of the circumference of a circle with centre A and radius 6m;

one third of the circumference of a circle with centre B and radius 5m;

one third of the circumference of a circle with centre C and radius 4m;

one third of the circumference of a circle with centre A and radius 3m;

one third of the circumference of a circle with centre B and radius 2m;

one third of the circumference of a circle with centre C and radius 1m;

Therefore the distance he walks is

$$\left(\tfrac{1}{2} \times 12\pi + \tfrac{1}{3} \times 10\pi + \tfrac{1}{3} \times 8\pi + \tfrac{1}{3} \times 6\pi + \tfrac{1}{3} \times 4\pi + \tfrac{1}{3} \times 2\pi\right)\text{m}$$

$$= \left(6\pi + \tfrac{1}{3} \times 30\pi\right)\text{m} = 16\pi\,\text{m}.$$

---

3. The coordinates of the midpoint of BC are (6, 3). The equation of the line of gradient 4 which passes through this point is $y = 4x - 21$.

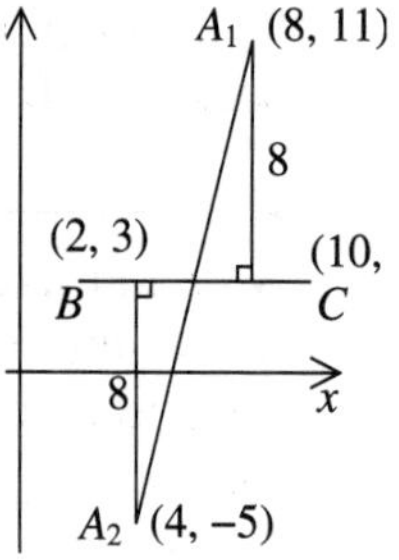

Now the length of BC is 8 and therefore the perpendicular height of triangle ABC must also be 8 as its area is 32.

Hence the y-coordinate of A is $3 + 8$ or $3 - 8$.

When $y = 11$, $4x - 21 = 11$ i.e. $4x = 32$ so $x = 8$.

When $y = -5$, $4x - 21 = -5$ i.e. $4x = 16$ so $x = 4$.

Hence the coordinates of the two possible positions of A are (8, 11) and (4, −5).

---

4. (a) As a fold along PQ places B on point D, PQ is the perpendicular bisector of BD. Hence PQ is parallel to AD and therefore triangles BPQ and BAD are similar. The ratio $PQ : AD = BQ : BD = 1 : 2$. By a similar argument, RS is the perpendicular bisector of CD and $RS : AD = 1 : 2$. Therefore PQ and RS are both perpendicular to BC and $PQ = RS = \frac{1}{2}AD$. Hence PR is parallel to BC and therefore perpendicular to AD.

We deduce that PR is the perpendicular bisector of AD and therefore a fold made along PR places vertex A on point D.

Note: the following alternative method uses the mid-point theorem to prove that PR is perpendicular to AD. A fold along PQ places B on D. Therefore PQ is the perpendicular bisector of BD and is parallel to AD. Hence $\angle BPQ = \angle PAD$ (corresponding angles).

Also $\angle BPQ$ folds onto $\angle DPQ$ so these are equal. Now $\angle DPQ = \angle PDA$ (alternate angles) and hence $\angle PAD = \angle PDA$. Thus triangle PAD is isosceles and so $PA = PD$.

Now PB folds onto PD so $PB = PD = PA$ and similarly $AR = RD = RC$. The midpoints of AB and AC are, therefore, P and R respectively and, from the mid-point theorem, PR is parallel to BC. Hence PR is perpendicular to AD.

Also, by applying the midpoint theorem to triangle ABD, we see that $PQ = \frac{1}{2}AD$ and so PR is the perpendicular bisector of AD. Therefore a fold made along PR places vertex A on point D.

(b) Let $BQ = x$. Then $BD = 2x$ [from (a)] and $CD = a - 2x$.

In right-angled triangle ABD: $AD^2 = AB^2 - BD^2 = c^2 - 4x^2$.

In right-angled triangle ACD: $AD^2 = AC^2 - CD^2 = b^2 - (a - 2x)^2$.

Therefore
$$\begin{aligned} c^2 - 4x^2 &= b^2 - (a - 2x)^2 \\ &= b^2 - \left(a^2 - 4ax + 4x^2\right) \\ &= b^2 - a^2 + 4ax - 4x^2 \end{aligned}$$

hence:
$$4ax = a^2 + c^2 - b^2$$

and therefore:
$$BQ = x = \frac{a^2 + c^2 - b^2}{4a}.$$

Note: a shorter method makes use of the Cosine Rule.

$$\cos\angle ABC = \frac{a^2 + c^2 - b^2}{2ac};$$

$$BQ = \tfrac{1}{2}BD = \tfrac{1}{2}AB \times \cos\angle ABC = \tfrac{1}{2}c \times \frac{a^2 + c^2 - b^2}{2ac} = \frac{a^2 + c^2 - b^2}{4a}.$$

---

5. *Note: in the manipulation of inequalities which follows, all the steps are valid because the numbers involved are* ***positive****.*

Observe that if $a \leqslant b$ and $a \geqslant 2$, then $4 \leqslant a+2 \leqslant b+2$ and so $\frac{1}{a+2} + \frac{1}{b+2} \leqslant \frac{1}{2}$. However, $\frac{1}{2} + \frac{1}{c+2} > \frac{1}{2}$ and we deduce that a must be 1.

Substituting $a = 1$ into the given equation gives

$$\frac{1}{3} + \frac{1}{b+2} = \frac{1}{2} + \frac{1}{c+2}, \text{ i.e. } \frac{1}{b+2} = \frac{1}{6} + \frac{1}{c+2}. \qquad (*)$$

From $(*)$: $\frac{1}{b+2} > \frac{1}{6}$, i.e. $b + 2 < 6$ so $b < 4$.

Since $b \geqslant a$ and b is an integer, we obtain $b = 1, 2$ or 3.

Substitute each value in turn into $(*)$:

b	1	2	3
$\frac{1}{c+2}$	$\frac{1}{3} - \frac{1}{6} = \frac{1}{6}$	$\frac{1}{4} - \frac{1}{6} = \frac{1}{12}$	$\frac{1}{5} - \frac{1}{6} = \frac{1}{30}$
c	4	10	28

Hence we obtain three ordered triples (a, b, c), namely, $(1, 1, 4)$, $(1, 2, 10)$ and $(1, 3, 28)$.

Note: an alternative method, having established that $a = 1$, is as follows.

Substituting for a in the original equation gives

$$\frac{1}{3} + \frac{1}{b+2} = \frac{1}{2} + \frac{1}{c+2} \quad \text{i.e.} \quad \frac{1}{b+2} - \frac{1}{c+2} = \frac{1}{6}.$$

Multiplying throughout by $6(b + 2)(c + 2)$ gives

$$\begin{aligned} (b+2)(c+2) &= 6(c+2) - 6(b+2) \\ bc + 2b + 2c + 4 &= 6c - 6b \\ bc + 8b - 4c - 32 &= -36 \\ \text{i.e.} \qquad (b-4)(c+8) &= -36. \end{aligned}$$

We deduce that, because $c + 8$ is positive, $b - 4$ is negative and hence $1 \leqslant b \leqslant 3$.

(NB this method is valid only because b and c are positive integers.)

When $b = 1$, $-3(c + 8) = -36$, i.e. $c + 8 = 12$, hence $c = 4$.

When $b = 2$, $-2(c + 8) = -36$, i.e. $c + 8 = 18$, hence $c = 10$.

When $b = 3$, $-(c + 8) = -36$, i.e. $c + 8 = 36$, hence $c = 28$.

This gives the same answer as above.

2000 Invitational Mathematics Challenge

(Grade 11)

TIME ALLOWED: 2 hours

Rulers and compasses may be used, **but not** protractors or squared paper

Calculators are not permitted. It is expected that all calculations and answers will be expressed as exact numbers such as 4π, $2 + \sqrt{7}$, etc. Marks are awarded for completeness, clarity, and style of presentation. A correct solution poorly presented will not earn full marks.

1. A horizontal wooden frame is made in the form of an equilateral triangle with each side 1 metre in length. Sean attaches a 6 metre rope to the frame at A and walks to point P so that P, A, and B lie on a straight line, as shown. Keeping the rope taut and horizontal at all times, Sean walks in the direction shown. How far does he walk if the rope gets wound twice round the frame?

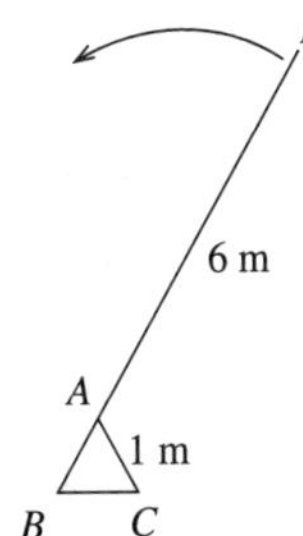

2. Solve the system of equations:
$$x^2 + xy + 2y^2 = 8$$
$$x^2 + y = 8.$$

3. The points B and C have coordinates $(-10, -10)$ and $(10, 10)$. Determine the coordinates of all points A such that A lies on the circle whose diameter is BC and such that the area of triangle ABC is 120.

4. Determine all ordered triples of positive integers (a, b, c) which satisfy
$$\frac{1}{a+2} + \frac{1}{b+2} = \frac{1}{2} + \frac{1}{c+2}.$$
[Ordered, here, means that $a \leqslant b \leqslant c$.]

5. (a) The incircle of a triangle is the circle which is tangent to each of the three sides of the triangle. In triangle ABC, the incircle has centre I and radius r. If the sides of the triangle have lengths a, b and c as shown, prove that $r = \dfrac{2\Delta}{a+b+c}$ where Δ is the area of triangle ABC.

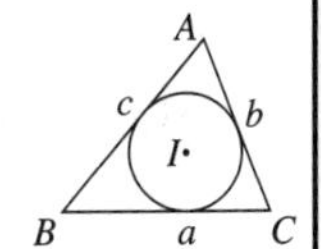

(b) In ΔABC, right-angled at B, a semicircle is drawn with its diameter on hypotenuse AC so that it is tangent to AB and BC. The radius of this semicircle is r_b. Two other semicircles are drawn so that their diameters lie on sides AB and BC. Each has one end of its diameter at vertex B and each is tangent to hypotenuse AC. The radii of these semicircles are r_c and r_a respectively. If the incircle of triangle ABC has radius r, prove that
$$\frac{2}{r} = \frac{1}{r_a} + \frac{1}{r_b} + \frac{1}{r_c}.$$

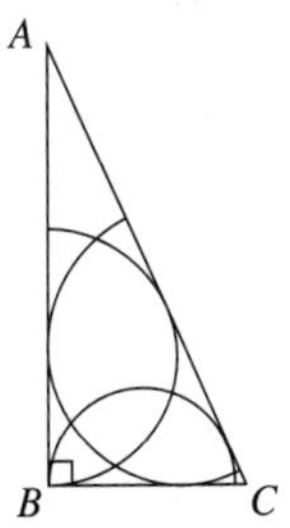

Canadian Mathematics Competition

An activity of The Centre for Education in Mathematics and Computing, Faculty of Mathematics, University of Waterloo, Waterloo, Ontario, Canada

Solutions for 2000 Invitational Mathematics Challenge (Grade 11)

---

1. Sean walks:
half of the circumference of a circle with centre A and radius 6m;
one third of the circumference of a circle with centre B and radius 5m;
one third of the circumference of a circle with centre C and radius 4m;
one third of the circumference of a circle with centre A and radius 3m;
one third of the circumference of a circle with centre B and radius 2m;
one third of the circumference of a circle with centre C and radius 1m;

Therefore the distance he walks is

$$\left(\tfrac{1}{2}\times 12\pi + \tfrac{1}{3}\times 10\pi + \tfrac{1}{3}\times 8\pi + \tfrac{1}{3}\times 6\pi + \tfrac{1}{3}\times 4\pi + \tfrac{1}{3}\times 2\pi\right)\text{m}$$

$$= \left(6\pi + \tfrac{1}{3}\times 30\pi\right)\text{m} = 16\pi\,\text{m}.$$

---

2. The given equations are

$$x^2 + xy + 2y^2 = 8 \ldots (1); \qquad x^2 + y = 8 \ldots (2).$$

Subtract (2) from (1) to obtain

$$xy + 2y^2 - y = 0$$

$$\text{i.e.} \quad y(x + 2y - 1) = 0$$

Hence either $y = 0$ or $x + 2y - 1 = 0$.
When $y = 0$, $x^2 = 8$, i.e. $x = \pm\sqrt{8}$.
When $x + 2y - 1 = 0$, $x = 1 - 2y$.

Substitute for x in (2) to obtain $(1 - 2y)^2 + y = 8$

$$\text{i.e.} \quad 1 - 4y + 4y^2 + y = 8$$

$$\text{i.e.} \quad 4y^2 - 3y - 7 = 0$$

$$\text{i.e.} \quad (4y - 7)(y + 1) = 0$$

$$\text{Thus} \quad y = \tfrac{7}{4} \quad \text{or} \quad -1.$$

When $y = \frac{7}{4}$, $x = 1 - \frac{7}{2} = -\frac{5}{2}$.
When $y = -1$, $x = 1 + 2 = 3$.
Thus there are four solution pairs (x, y), namely, $(\sqrt{8},0)$, $(-\sqrt{8},0)$, $\left(-\frac{5}{2}, \frac{7}{4}\right)$ and $(3, -1)$.

---

3. The centre of the circle is the midpoint of BC, namely the origin, O. The radius of the circle is the length of OC which is $\sqrt{200}$, i.e. $10\sqrt{2}$.

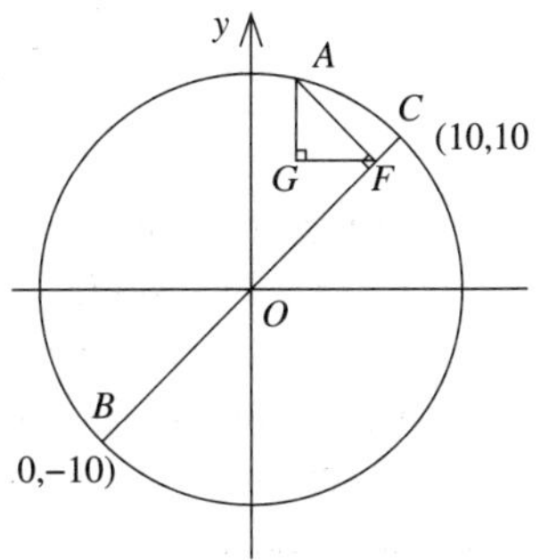

Taking BC to be the base of triangle ABC, the height, h, of the triangle is given by the equation $\frac{1}{2} \times 20\sqrt{2} \times h = 120$, i.e. $h = 6\sqrt{2}$.

Let F be the foot of the perpendicular from A to BC.

Then, in right-angled triangle OAF:

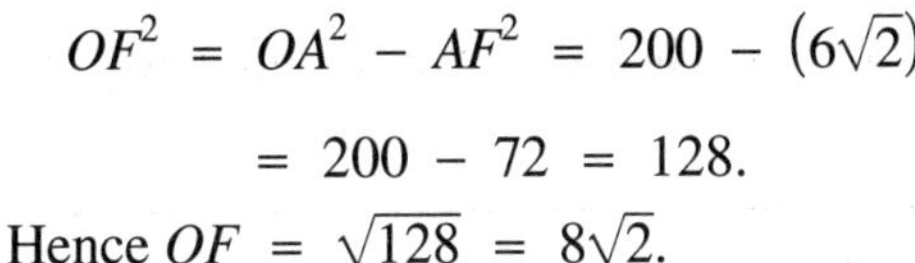

$$OF^2 = OA^2 - AF^2 = 200 - \left(6\sqrt{2}\right)^2$$

$$= 200 - 72 = 128.$$

Hence $OF = \sqrt{128} = 8\sqrt{2}$.

F lies either on OB or OC. Consider the case in which F lies on OC and A is above the line BC. Then, as F lies on the line whose equation is $y = x$ and $OF = 8\sqrt{2}$, F is the point (8, 8).Now let G be the point such that AG is vertical and GF is horizontal.

Line FA has gradient -1 since it is perpendicular to line BC and hence triangle AFG is an isosceles right-angled triangle in which $FG = GA$ and $AF = 6\sqrt{2}$.

We deduce that $FG = GA = 6$ and hence A is the point $(8 - 6,\ 8 + 6)$ i.e. (2, 14).

By an identical argument, when F lies on OC and A is below the line BC, we find that A is the point $(8 + 6,\ 8 - 6)$ i.e. (14, 2).

The other two possible positions of A are the reflections of these two points in the line whose equation is $y = -x$. Hence the possible positions of A are (2, 14), (14, 2), (−14, −2) and $(-2, -14)$.

---

4. *Note: in the manipulation of inequalities which follows, all the steps are valid because the numbers involved are **positive**.*

Observe that if $a \leqslant b$ and $a \geqslant 2$, then $4 \leqslant a + 2 \leqslant b + 2$ and so $\frac{1}{a + 2} + \frac{1}{b + 2} \leqslant \frac{1}{2}$. However, $\frac{1}{2} + \frac{1}{c + 2} > \frac{1}{2}$ and we deduce that a must be 1.

Substituting $a = 1$ into the given equation gives

$$\frac{1}{3} + \frac{1}{b + 2} = \frac{1}{2} + \frac{1}{c + 2}, \text{ i.e. } \frac{1}{b + 2} = \frac{1}{6} + \frac{1}{c + 2}. \qquad (*)$$

From (∗): $\frac{1}{b + 2} > \frac{1}{6}$, i.e. $b + 2 < 6$ so $b < 4$.

Since $b \geqslant a$ and b is an integer, we obtain $b = 1, 2$ or 3.

Substitute each value in turn into (∗):

b	1	2	3
$\frac{1}{c+2}$	$\frac{1}{3} - \frac{1}{6} = \frac{1}{6}$	$\frac{1}{4} - \frac{1}{6} = \frac{1}{12}$	$\frac{1}{5} - \frac{1}{6} = \frac{1}{30}$
c	4	10	28

Hence we obtain three ordered triples (a, b, c), namely, $(1, 1, 4)$, $(1, 2, 10)$ and $(1, 3, 28)$.

Note: an alternative method, is shown in the solution to Grade 10 question 5.

---

5. (a) The area of triangle $BIC = \frac{1}{2}ar$; the area of triangle $AIC = \frac{1}{2}br$ and the area of triangle $AIB = \frac{1}{2}cr$. Thus, Δ, the area of triangle ABC is given by

$$\Delta = \tfrac{1}{2}ar + \tfrac{1}{2}br + \tfrac{1}{2}cr$$

$$= \tfrac{1}{2}(a + b + c)r.$$

Hence $r = \dfrac{2\Delta}{a + b + c}$.

(b) Consider the semicircle which touches AC and BC, i.e. the semicircle whose radius is r_c. Let the centre of this semicircle be D and join D to C.

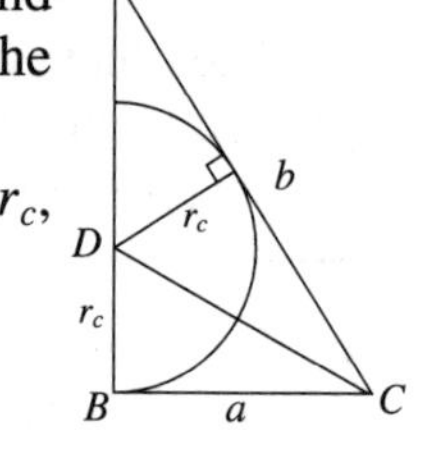

Now triangle DBC has base a and height r_c, whilst triangle DAC has base b and height r_c.

Area of triangle ABC

= area of triangle DBC + area of triangle DAC

i.e. $\quad \Delta = \tfrac{1}{2}ar_c + \tfrac{1}{2}br_c = \tfrac{1}{2}(a + b)r_c$

i.e. $\quad r_c = \dfrac{2\Delta}{a + b}$.

Repeating this procedure for the other two semicircles gives:

$$r_a = \frac{2\Delta}{b + c} \quad \text{and} \quad r_b = \frac{2\Delta}{c + a}.$$

Hence: $\quad \dfrac{1}{r_a} + \dfrac{1}{r_b} + \dfrac{1}{r_c} = \dfrac{b + c + c + a + a + b}{2\Delta}$

$$= \frac{2(a + b + c)}{2\Delta} = \frac{2}{r} \text{ [from (a)]}.$$

Comments on IIIMC Question Papers and Scripts:

The marking team gathered on Friday 19 May at The Village, Leeds. After a pleasant evening's organisation of marks sheets, all retired for a good night's sleep before the marking commenced.

There was much to be pleased about in the scripts of the participants this year, The questions which were in common to both papers were well attempted. However, there are concerns that pupils have little appreciation of formal proof and, as happened before, geometry questions were poorly done. However, there are signs that pupils can work reasonably well with coordinate geometry. One thing which was immediately obvious was pupils' reluctance to consider the existence of a second solution when taking a square root. There was much confusion in fractions work with a number of pupils equating $\frac{1}{a+2}$ to $\frac{1}{a} + \frac{1}{2}$.

IIIMC awards and high scorers

As in the JMO and the Kangaroo, everyone is awarded a certificate with about a quarter obtaining a Certificate of Distinction (a score of 19 in IIIMC 10 and a score of 22 in IIIMC 11) and the rest a Certificate of Participation. In addition, the top fifty in the Year 10 competition were each given a copy of the book *Mathematics for the Curious* by Peter Higgins and the top fifty in the Year 11 competition were each given a copy of the book *More Joy of Mathematics* by Theoni Pappus.

The whole of the UKMT hopes that pupils and teachers found this a rewarding experience and look forward to further participation in future.

Higher scorers were as follows

IIIMC 10

47	Nathan Bowler	Knutsford HS
44	Thomas Smith	Latymer Uppers
41	Paul Jefferys	Berkhamsted Collegiate
40	Rui Huo	Royal Russell School
	Thomas Rosoman	Kings School Worcester

39	James Burberry	Queen Elizabeths Hosp
	Michael Chan	Bedford School
	Rose Chen	Queen Anne School
	Tom Williams	Fortismere
38	Inkwon Choi	Yarm School
37	Jenny Gardner	Tiffin Girls GS
	Bryn Garrod	Ke VI Camp Hill Boys School
	Alex Smith	Ke VI Five Ways School
36	Gavin Johnstone	Dame Alice Owens School
	Yuan Shen	George Watsons College
35	Ying Lu	Royal Russell School
34	Ruari Kerr	KES Birmingham
	Edwin Lang	Chigwell School
	Jessica Owens	Guildford High School
33	Peter Conlon	Worth School
	Adam J P Dewbery	Kings College Taunton
	Peyman Owladi	Latymer Upper School
	Johnny Tang	Wilsons
32	Hyeyoun Chung	St Pauls Girls School
	Martyn Jones	RGS Newcastle
31	Yea lin Chung	Arden
	Paul Gilbert	Westminster School
	Joanna Harbour	Dame Alice Harpur
	Dominic Weinberg	Dr Challoners Grammar School
30	Chris Clarke	Methodist College
	James Griffin	RGS Worcester
	Claire Jones	Tonbridge GS for Girls
	Katherin Laidlar	Withington Girls School
	Kirjusha Makharinsky	Dr Challoners Grammar School
	Erica Thompson	Douglas Academy
	Tom Ward	St Pauls London

29	Matthew Austin	Oundle School
	Edward Heaney	Abingdon School
	Yoona Lee	Chiswick Community CS
	Thomas Pickup	Westminster School
	James Robinson	Winchester College
	Jacob Shepherd	Winchester College
	Chris Thomas	KE VI Camp Hill Boys School
28	Ben Crowne	City of London School
	Alastair Currie	George Watsons College
	Ian Elder	Conyers
	Katie Gough	Alleyns School
	Edmund Naylor	Westminster School
	Amanda Price	Croham Hurst School

IIMC 11

45	Stephen Burgess	Shrewsbury School
	Andrius Simenas	Ampleforth College
43	Tom Niu	Bromsgrove School
41	James Orme	Tonbridge School
40	Timothy Austin	Colchester RGS
39	George Walker	Shrewsbury School
38	Dustin Cheng	Framlingham College
37	David Wyatt	Winchester College
35	Jameel Kassam	Merchant Taylors School
	Rachel Price	Haberdasher Aske's School for Girls
	Thomas West	Oundle School
33	Neil Davidson	RGS Guildford
	Timothy Northover	Queen Elizabeths Hosp
32	Victoria Neale	Henry Beaufort School
	Timothy Ward	Shrewsbury School
	Matthew Wright	Bradford Grammar School

30	Richard Paxton	QEGS Wakefield
	Charles Roome	RGS Guildford
29	Charles Barrow	Charterhouse
	Adam Biltcliffe	High Storrs School
	Yevgen Nikulin	Eastbourne College
	Jeremy Reizenstein	St Pauls London
	Chris Taylor	Bryanston School
	Paween Thanarat	Shrewsbury School
28	Adam Edelshain	St Pauls London
	James Hood	Wellington College
	Andrew Huang	KES Birmingham
	Richard McDowell	Clifton College
	Colin Reid	Loughborough Grammar
27	Peter Batten	Poltair School
	Nicholas Golding	Warwick School
	Ian Webb	Chesham High School
	Jack Willis	Churston Ferrers GS
26	Alistair Bird	Tonbridge School
	James Bland	Winchester College
	Edmund Bolton	Bishop Wordsworths
	Jonathan Middleton	Nottingham HS
	Colin Palmer	Warwick School
25	Ross Beaton	Methodist College
	Joseph Catling	Christleton High School
	Grace Chen	High Storrs School
	David Stuckey	KE VI Camp Hill Boys School
	Toby Wood	Westcliff High School for Boys
	Alex Wright	Oundle School

A certificate from the IIIMC

2000 Invitational Mathematics Challenge

....

of

....

received a

CERTIFICATE of DISTINCTION

Canadian Mathematics Competition

Peter Neumann

Chairman, United Kingdom Mathematics Trust

THE INTERMEDIATE INTERNATIONAL INVITATIONAL MATHEMATICAL CHALLENGE

The Challenge is open only to the best candidates from Year 11 and below, judged by their achievement in the UK Intermediate Mathematical Challenge. Only 600 candidates are invited to sit each of the Year 10 and Year 11 papers. It is also taken by candidates in many other countries.

It is a two hour paper requiring fully detailed solutions to five demanding problems which must be tackled without a calculator. The knowledge and skills required are often beyond the range of normal school work.

National Mathematics Summer School

The seventh annual National Mathematics Summer School was held at Queen's College, Birmingham during the period 3-7 July 2000. The chief organiser was Professor Adam McBride (University of Strathclyde). Throughout the week a great deal of valuable assistance was provided by Mr Richard Atkins (Oundle School).

Selection for the Summer School is based on performance in the IIIMC. Around 40 students from Y10 and Y11 are invited each year, with one or two younger students occasionally included too. In addition, the IMO team are in attendance for the final stages of their training. Although they are engaged in separate activities for much of the time, they join the younger students for the team competitions during the second half of the morning and also for social events in the evening.

For completeness, a full list of participants and a programme are supplied below. However, it will help to convey the spirit of the event if aspects of the programme are described in more detail.

Things got under way on the evening of Monday 3 July. The opening lecture was given by Professor Chris Robson (University of Leeds), who showed us that he is an expert in two different types of Ring Theory. One is the area of abstract algebra which bears that name. The other, which provided the basis for the lecture, is bell-ringing. With a set of eight handbells available, there was scope for lots of audience participation. The lecture served as an excellent ice-breaker and got proceedings off to a fine start.

The format on the following days was the same each time. After breakfast there was a series of four linked 'masterclasses' led this year by Mr Nick Lord (Tonbridge School). The students were exposed to a whole range of beautiful problems, featuring such joys as the Euler line and the Nine-Point Circle. There were plenty more problems for the students to tackle themselves and in the final session they were invited to present their solutions on the OHP.

After refreshments, it was time for team competitions, superbly organised as last year by Richard Atkins. To get the full flavour you really had to be there but here is an attempt to convey some of what went on. We had 6 teams, each consisting of an IMO team member as 'captain' and 6 or 7 of the younger students. Some of the latter were deemed 'juniors' and the rest 'seniors'. Each team was split into two groups who sat at separate tables. In the first round, each student tackled a set of three problems, with different sets for the juniors and seniors. Students worked individually without conferring. Answers only were required. When time was up, these answers were taken away for marking by the staff (who had also been beavering away on the problems to check the answers!). Meanwhile the second round got under way. Each team got 8 problems which were to be shared out, one per student. The 'captain' had to decide on the optimum allocation. This time, full solutions were to be written out. This took about 25-30 minutes. Finally came the relays. Each relay contained four problems but later problems could only be attempted once the answer to the previous question had been received. Each team was divided into four 'pairs', each pair doing one leg of the relay. If the first pair passed on the wrong answer to the next pair, everything went haywire. Again, captains had to decide on who should be in the first pair to minimise the risk of disaster. Things proceeded at a frantic pace, but fortunately no physical injuries were reported. Points accumulated over the four days and the teams proved to be pretty evenly matched. In summary, each day we had almost two hours of all-action stuff and lots of fun.

For the afternoon sessions the youngsters were split into two groups of equal size. During the four days they covered seven different topics. The novelties this year were the sessions on Logical Puzzles by Dan Crisan, Combinatorial Geometry by Ben Green and Games of Strategy by Howard Groves. In each session, the aim was to get the students working on problems, with a minimum of exposition.

Two events have become an integral part of the Summer School. One is a trip to the theatre. This year we went to the Birmingham Rep to see 'Present Laughter' by Noel Coward. The production was set in the 1930s and reflected the culture and *mores* of the day. Nevertheless, 60 years on, it wasn't hard for even the younger members of the audience to follow the plot. However, there is more to the theatre trip than the actual show. There is the journey to the theatre and back. The tradition has evolved that we walk, in teams with the captain at the head. Our route took us along the towpath by a canal. The sky was dark but we escaped with just the odd spot of rain. (We found out the next day how lucky we had been; not far away there had been flash floods with water inches deep in some streets.) By this stage, the students had got to know each other and many interesting conversations took place.

The other item without which no Summer School would be complete is the 'Musical Evening'. Students had been invited to bring along suitable party pieces and we were treated to a varied selection, mostly musical rather than literary but also including some assorted card tricks.

Things drew to a close around 15.30 on Friday afternoon and we all went our separate ways. Everyone was pretty tired but the Summer School had succeeded in its aims. A group of talented young mathematicians had been exposed to a wide range of activities intended to stimulate and stretch them. They have plenty of material to study in the weeks ahead. It is to be hoped that they will enter the fray again next year. Perhaps one or two could find themselves in the IMO team for 2001.

Finally, thanks are due to all who gave unstintingly of their time to make the event the undoubted success it was. Particular mention should be made of Nicole Mainwaring and the domestic staff at Queen's College who made us most welcome and catered so well for all our needs.

As well as the 'official story' above, here is the view of one of the participants.

As I had never heard of this event before being invited to go, it was a rather daunting prospect, especially as I would have to make the train journey to Birmingham alone. However, I arrived intact, and managed to find my room. I then started talking to some other people; the girls 'teamed up' quite rapidly, and we chatted over dinner, albeit rather nervously. It soon became apparent that the boys needed some help with their social skills, so we mixed up a bit for the first lecture, 'Sound as a Bell'. This was very interesting, not least because none of us knew anything about campanology whatsoever. However, by the end of the lecture lots of us had rung some bells (some at the right time!), and done some maths as well.

We soon found the common room, with its attractions of a pool table and some packs of cards and, by the end of the week, most people were chatting and playing card games or pool each evening, and during breaks.

The format of each day was the same: breakfast (which most people managed to get up for), followed by a geometry lecture given by Nick Lord. These were very interesting, although hard work for some of us as we had never encountered geometry like this before (i.e. it was not on the GCSE syllabus). Three of the four sessions consisted of being taught new things, having proofs demonstrated to us, and then working on some problems; the fourth gave us an opportunity to show our solutions to these problems to the rest of the group. After this came a break, complete with biscuits!

After the break came the team games. For these we were split into six groups (named after mathematicians of course), with members of the IMO squad as captains. Firstly we worked on individual problems, requiring only an answer. We then had 'Team choice': a selection of eight varying problems, each requiring a full solution. Our captains allocated these problems, as only one solution per question would count so some people would have to do the hard ones! This was followed by the relays. Each team was divided into four pairs, who each had a problem to work on.

However, the second problem required the solution to the first, the third the solution from the second, and so on, so they had to be done in order. These caused lots of frowns, because obviously if one solution was incorrect, all the rest would be! Points were awarded for each answer, with bonus points for the teams coming in the first three. The points for the team games were totalled throughout the week, and it was actually very close all the way through.

After lunch (another necessity) were more lectures, but this time each one was given to half the group at a time. These were on various subjects, including Diophantine equations, inequalities, combinatorial geometry and game strategies, and all were interesting, especially as they covered a wide range of subjects. Quite a few of us had never even heard of some of the topics, let alone learnt anything about them previously! We had two of these sessions each afternoon, with yet more biscuits in between!

After dinner there were a range of social activities: one evening we watched a video about Paul Erdös, another we put on a musical evening with a range of contributions, from Mussorgsky to Tom Lehrer. One young lady also performed some highly complex card tricks with the assistance of her friend; they baffled us (nearly) all evening. The third evening we went to the theatre in Birmingham to see *Present Laughter*, by Noel Coward. I think for many of us this was the first time we had seen any Coward, and it was an enjoyable evening. The walk along the towpath to and from the theatre provided plenty of opportunities for conversation, and thankfully it did not rain much, if at all.

At the end of the week we all exchanged e-mail addresses and, thanks to the wonders of modern technology, we can now keep in touch with each other on a regular basis.

I think we are all very grateful to the people who organised and ran this summer school, especially Adam McBride, without whom it would not have run so smoothly. We all went away with plenty of things to think about, and a realisation that there is more to Mathematics than most of us had previously realised.

UKMT Summer School 2000 Programme

Monday 3 July

17.00 onwards	Arrival and Dinner
19.30-21.00	Welcome and Opening Lecture – Chris Robson: *Sound as a Bell*

Tuesday 4 July

09.00-10.30	Geometry Lecture 1
11.00-12.45	Team Competitions
14.00-15.45	Small groups
16.00-17.45	Small groups
18.30	Dinner followed by mathematical videos

Wednesday 5 July

09.00-10.30	Geometry Lecture 2
11.00-12.45	Team Competitions
14.00-15.45	Small groups
16.00-17.45	Small groups
18.30	Dinner followed by a 'musical evening'

Thursday 6 July

09.00-10.30	Geometry Lecture 3
11.00-12.45	Team Competitions
14.00-15.45	Small groups
16.00-17.45	Small groups
18.00	Dinner then theatre: *Present Laughter* by Noel Coward

Friday 7 July

09.00-10.30	Geometry Lecture 4
11.00-12.45	Team Competitions
14.00-15.45	Small groups
15.45	Tea and depart

Staff and topics:

Richard Atkins – Inequalities
Christopher Bradley – Sequences
Dan Crisan – Logical Puzzles
Julian Gilbey – Catalan Numbers
Ben Green – Combinatorial Geometry
Howard Groves – Games of Strategy
Nick Lord – Geometry
Adam McBride – Diophantine Equations

Pupils who attended were

Peter Batten	Poltair Community School
Nathan Bowler	Knutsford HS
James Burberry	QEH, Bristol
Alison Campbell Smith	Sutton HS
Michael Chan	Bedford School
Peter Conlon	Worth School
Adam Dewbery	Kings College, Taunton
Adam Edelshain	St. Paul's School
Jenny Gardner	Tiffin Girls
Bryn Garrod	KE VI Camp Hill Boys
Nicholas Golding	Warwick School
James Griffin	RGS, Worcester
Joanna Harbour	Dame Alice Harpur School
Gavin Johnstone	Dame Alice Owen's School
Claire Jones	Tonbridge Girls
Jameel Kassam	Merchant Taylors School
Ruari Kerr	KES, Birmingham
Katherin Laidlar	Withington Girls
Edwin Lang	Chigwell School
Vicky Neale	Henry Beaufort School
Yevgen Nikulin	Eastbourne College
James Orme	Tonbridge School
Jessica Owens	Guildford HS
Richard Paxton	QEGS, Wakefield
Rachel Price	Haberdashers' Aske's Girls
Thomas Rosoman	Kings School, Worcester
Yuan Shen	George Watson's College
Alex Smith	KE VI Five Ways
Chris Taylor	Bryanston School
Chris Thomas	KE VI Camp Hill Boys
Erica Thompson	Douglas Academy
Timothy Ward	Shrewsbury School
Ian Webb	Chesham HS
Dominic Weinberg	Dr. Challoner's
Tom Williams	Fortismere School
Matthew Wright	Bradford GS

IMO Squad

Thomas Barnet-Lamb	Westminster School
Stephen Brooks	Abingdon School
David Collier	KE VI Southampton
David Knipe	Sullivan Upper School
Michael Spencer	Lawnswood HS, Leeds
Oliver Thomas	Winchester College

Staff

Richard Atkins	Oundle School
Christopher Bradley	Clifton College, Bristol
Dan Crisan	Queen's College, Cambridge
Julian Gilbey	QMW, University of London
Ben Green	Trinity College, Cambridge
Howard Groves	RGS, Worcester
Imre Leader	University College, London
Nick Lord	Tonbridge School
Adam McBride	University of Strathclyde, Glasgow
Chris Robson	University of Leeds

The Senior Mathematical Challenge and British Mathematical Olympiads

The Senior Challenge took place on Tuesday 9th November 1999. Some 32,211 pupils took part and around 1,000 took part in the next stage – British Mathematical Olympiad Round 1 on Wednesday, 12 January 2000.

UK SENIOR MATHEMATICAL CHALLENGE

Tuesday 9 November 1999

Organised by the **United Kingdom Mathematics Trust**

RULES AND GUIDELINES (to be read before starting)

1. Do not open the question paper until the invigilator tells you to do so.
2. Detach the Answer Sheet (back page) and fill in your personal details before you open the question paper and begin.
   Once you have begun, record all your answers on the Answer Sheet.
3. Time allowed: **90 minutes**.
   No answers or personal details may be entered on the Answer Sheet after the 90 minutes are over.
4. The use of rough paper is allowed.
   Calculators, measuring instruments and squared paper are forbidden.
5. Candidates must be full-time students at secondary school or FE college, and must be in Year 13 or below (England & Wales); S6 or below (Scotland); Year 14 or below (Northern Ireland).
6. There are twenty-five questions. Each question is followed by five options marked *A*, *B*, *C*, *D*, *E*. Only one of these is correct. Enter the letter *A*-*E* corresponding to the correct answer in the corresponding box on the Answer Sheet.
7. **Scoring rules**: all candidates start out with 25 marks;
   0 marks are awarded for each question left unanswered;
   4 marks are awarded for each correct answer;
   1 mark is deducted for each incorrect answer.
8. **Guessing**: Remember that there is a penalty for wrong answers. Note also that later questions are deliberately intended to be harder than earlier questions. You are thus advised to concentrate first on solving as many as possible of the first 15-20 questions. Only then should you try later questions.

1. How many prime numbers are there less than 20?

   A 6 *B* 7 *C* 8 *D* 9 *E* 10

2. What is the largest number of Sundays that there can be in any one year?

   A 50 *B* 51 *C* 52 *D* 53 *E* 54

3. Which of the following is not the net of a pyramid?

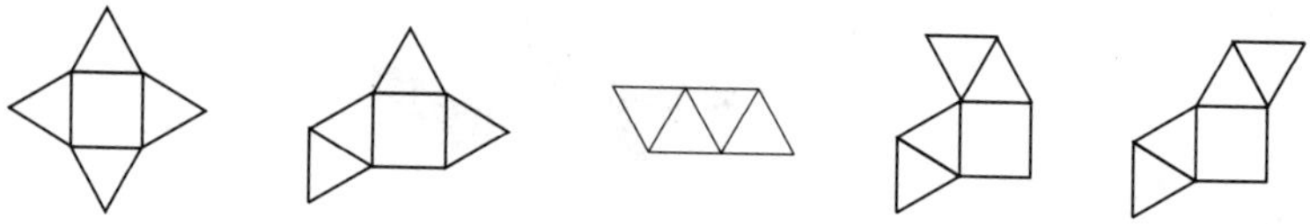

   A *B* *C* *D* *E*

4. I need to buy 12 films for my camera before my holiday. They normally cost £4.50 each, but a number of shops have "special offers". Which of these is the best deal?

   A One fifth off all prices! *B* Two for the price of four! *C* Buy two – get one free!
   D 30% price cut! *E* Pay only three quarters of the normal price!

5. In 1998 a newspaper reported that "The world record for remembering the value of π to the greatest number of decimal places is 40 000 places, which took the record holder 17 hours and 21 minutes to recite."

   What was the average number of decimal places recited per minute, approximately?

   A 20 *B* 40 *C* 200 *D* 400 *E* 2000

6. Our ancient Ancient History teacher's copy of Homer's *Odyssey* cost 40p in 1974. A similar edition today costs £5. What percentage increase is this?

   A 12.5% *B* 1150% *C* 1250% *D* 12400% *E* 12500%

7. The size of each exterior angle of a regular polygon is one quarter of the size of an interior angle. How many sides does the polygon have?

   A 6 *B* 8 *C* 9 *D* 10 *E* 12

8. Two numbers differ by 9 and have sum 99. What is the ratio of the larger number to the smaller?

   A 5:4 *B* 6:5 *C* 7:6 *D* 8:7 *E* 9:8

9. The factorial of n, written $n!$, is defined by $n! = 1 \times 2 \times 3 \times \ldots \times (n-2) \times (n-1) \times n$
   e.g. $6! = 1 \times 2 \times 3 \times 4 \times 5 \times 6 = 720$.

   What is the smallest positive integer which is *not* a factor of 50! ?

   A 51 *B* 52 *C* 53 *D* 54 *E* 55

10. Which is the largest of the following?

    A $\sqrt{1999}$ *B* $1\sqrt{999}$ *C* $19\sqrt{99}$ *D* $199\sqrt{9}$ *E* $1999\sqrt{0}$

11. In how many different ways can I circle letters in the grid shown so that there is exactly one circled letter in each row and exactly one circled letter in each column?

A	B	C	D	E
F	G	H	I	J
K	L	M	N	O
P	Q	R	S	T
U	V	W	X	Y

A 15 *B* 24 *C* 60 *D* 100 *E* 120

12. Earlier this year, the White Rabbit said to me, "Two days ago, Alice was still thirteen, but her sixteenth birthday will be next year." When is Alice's birthday?

A Jan 1st *B* Feb 28th *C* Feb 29th *D* Dec 30th *E* Dec 31st

13. Two square pieces of card, each 3 cm × 3 cm, are attached by a single pin to a board. The pin passes through a point 1/3 of the way along the diagonal of each square and the squares overlap exactly. The bottom card now remains fixed, while the top card is rotated through 180°. What is the area of overlap of the cards in this new position?

A 1 cm^2 *B* 2 cm^2 *C* 4 cm^2 *D* 6 cm^2 *E* 9 cm^2

14. The line whose equation is $y = 3x + 4$ is reflected in the line whose equation is $y = -x$. What is the equation of the image line?

A $3y = x + 4$ *B* $3y = x - 4$ *C* $y = 3x - 4$ *D* $y = -3x - 4$ *E* $y = 4x + 3$

15. Three people each think of a number which is the product of two different primes. Which of the following could be the product of the three numbers which are thought of?

A 120 *B* 144 *C* 240 *D* 3000 *E* 12100

16. When rounded to 3 significant figures, the number x is written as 1000. What is the largest range of possible values of x?

A $999 \leqslant x < 1001$ *B* $995 \leqslant x < 1005$ *C* $990 \leqslant x < 1010$
D $999.5 \leqslant x < 1005$ *E* $999.5 \leqslant x < 1000.5$

17. The ratio of Jon's age to Jan's age is 3 : 1. Three years ago the ratio was 4 : 1. In how many years time will the ratio be 2 : 1?

A 3 *B* 6 *C* 9 *D* 12 *E* 15

18. The diagram shows two concentric circles. The chord of the large circle is a tangent to the small circle and has length $2p$. What is the area of the shaded region?

A πp^2 *B* $2\pi p^2$ *C* $3\pi p^2$ *D* $4\pi p^2$
E more information needed

19. P is a vertex of a cuboid and Q, R and S are three points on the edges as shown. $PQ = 2$ cm, $PR = 2$ cm and $PS = 1$ cm. What is the area, in cm^2, of triangle QRS?

A $\sqrt{15}/4$ *B* $5/2$ *C* $\sqrt{6}$ *D* $2\sqrt{2}$ *E* $\sqrt{10}$

20. What is the 1999th term of the sequence 1, 2, 2, 3, 3, 3, 4, 4, 4, 4, 5, … ?

A 59 *B* 60 *C* 61 *D* 62 *E* 63

21. Just one of the following is a prime number. Which one is it?

A $1000^2 + 111^2$ *B* $555^2 + 666^2$ *C* $2000^2 - 999^2$
D $1001^2 + 1002^2$ *E* $1001^2 + 1003^2$

22.

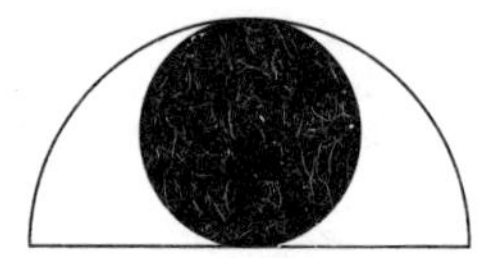 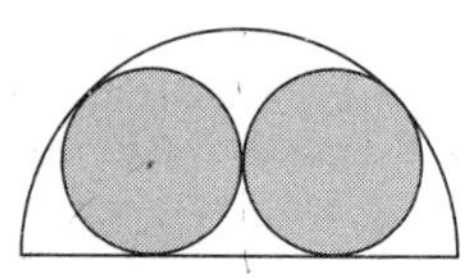

The area of each large semicircle is 2. What is the difference between the black and grey shaded areas?

A 0 *B* $\frac{1}{2}$ *C* $1 + 2\sqrt{2}$ *D* $\frac{5}{9}$ *E* $23 - 16\sqrt{2}$

23. The statement "There are exactly four integer values of n for which $(2n + y)/(n - 2)$ is itself an integer" is true for certain values of y only. For how many values of y in the range $1 \leqslant y \leqslant 20$ is the statement true?

A 0 *B* 7 *C* 8 *D* 10 *E* 20

24. The figure shows a hexagon *AZBXCY* made from four congruent tiles. The shape and position of the tiles are given by triangle *ABC* and the three reflections of triangle *ABC* in the lines determined by its sides. For example, *ABZ* is the image of *ABC* when reflected in the line determined by *AB*. If a polygon is made from five tiles whose shape and position are determined by a quadrilateral and the four reflections of that quadrilateral in the lines determined by its sides, what is the smallest possible number of sides of the resulting polygon?

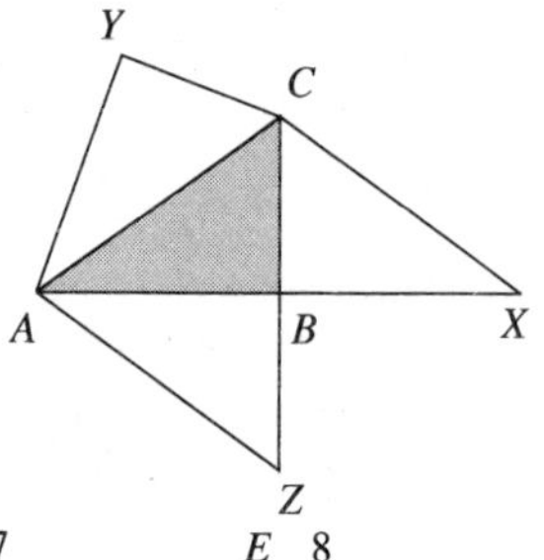

A 4 *B* 5 *C* 6 *D* 7 *E* 8

25. What is the sum to infinity of the convergent series

$$\frac{1}{2} + \frac{1}{4} + \frac{2}{8} + \frac{3}{16} + \frac{5}{32} + \frac{8}{64} + \frac{13}{128} + \frac{21}{256} + \frac{34}{512} + \ldots ?$$

A $\frac{7}{4}$ *B* 2 *C* $\sqrt{5}$ *D* $\frac{9}{4}$ *E* $\frac{7}{3}$

Futher remarks

Although designed as a multiple-choice paper, the Senior Challenge is different from the Junior and Intermediate in that it is marked in centres and a summary of the scores obtained sent to the Maths Challenge Office. Here they are collated to produce the cut-off scores for the certificates. To assist centres, a pupil answer sheet is incorporated into the question paper and this and the background information supplied are shown below.

UK SENIOR MATHEMATICAL CHALLENGE

TUESDAY 9 NOVEMBER 2000

ANSWER SHEET

1.	
2.	
3.	
4.	
5.	
6.	
7.	
8.	
9.	
10.	
11.	
12.	
13.	
14.	
15.	
16.	
17.	
18.	
19.	
20.	
21.	
22.	
23.	
24.	
25.	

To be completed by the student

SCHOOL/COLLEGE NAME ....

UKMT CENTRE NUMBER ....

YOUR NAME ....

SCHOOL YEAR ....

MATHS SET/ TEACHER ....

Enter the option (A, B, C, D or E) which corresponds to the correct answer for each question in the box for that question.

To be completed by the teacher not by the student

Each question is worth *four* marks.

One mark is deducted for each wrong answer.

No marks are deducted for questions left unanswered.

The total score is calculated by taking four times the number of correct answers, subtracting the number of wrong answers, and then adding 25.

Number correct [ ] × 4 = [ ]

– Number wrong [ ]

Difference [ ]

\+ 25

Total score [ ]

NB: Blank papers score 25.

INFORMATION

National organisation of the UK SMC

This year's UK Senior Mathematical Challenge is run under the auspices of the

United Kingdom Mathematics Trust (UKMT).

The UKMT committee responsible for the administration of the UK SMC is called the *Senior Challenge Subtrust* (SCS). Any comments, questions or suggestions you may have about the UK SMC should be sent to the Chair of that committee at the following address (if you need a reply, please enclose an SAE):

Bill Richardson, SCS, Elgin Academy, Morriston Road, Elgin, Morayshire IV30 4ND.

Accessibility

Most of the problems on the UK SMC are meant to be *accessible* – in the sense that all participants should understand them and want to solve them. However, this does not mean that the problems are easy: they are designed to make students *think* – sometimes in unfamiliar ways. In particular, candidates should be careful when interpreting their scores: a score of 60 or 70 may be much more creditable than might at first appear.

The UK has three different 'national curricula' in mathematics! This makes it hard to please everyone. Nevertheless we try to ensure that the first 15-20 UK SMC problems require little beyond what should be taught (even if it is not assessed!) in a good GCSE or Standard Grade mathematics course.

Organisation within schools and colleges

The paper must be taken on the official date.

Candidates must be invigilated strictly: they must be at separate desks, well separated, and all facing the same way.

In each school or college, the person named as 'contact teacher' on the Entry Form is responsible for ensuring that the paper is administered correctly.

Return of results

Immediately after the event, each centre must complete the machine-readable Results Sheet and return it, folded once only, in the reply-paid envelope provided, to:

UK SMC, School of Mathematics, University of Leeds, Leeds LS2 9JT.

The timetable for processing UK SMC results and for packing and returning certificates is tight. It is crucial that all results arrive in Leeds by Friday 12th November.

Certificates

The top 6-7% of candidates receive Gold certificates, the next 12-14% receive Silver certificates, and the next 18-21% Bronze.

The British Mathematical Olympiad

Round 1 of the BMO takes place on *Wednesday 12th January 2000*. All schools and colleges with high scoring UK SMC candidates automatically receive a BMO Entry Form with their UK SMC results. Entry is not restricted to these candidates; other centres may obtain an Entry Form by sending a stamped self-addressed envelope to:

Dr Alan West (BMO), School of Mathematics, University of Leeds, Leeds LS2 9JT.

Enquiries about the BMO should *not* be sent to the UK SMC organisers.

The solutions are provided in a leaflet which is also set up to facilitate the marking.

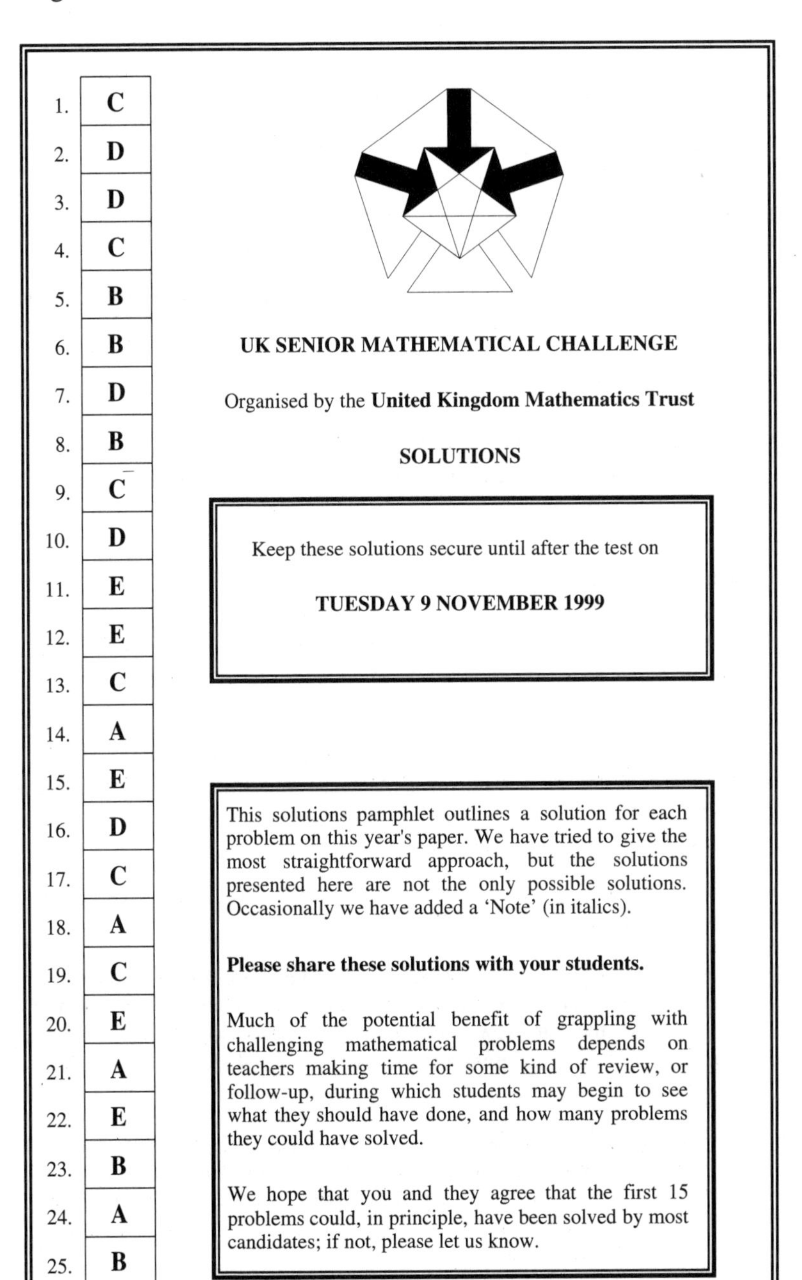

1.	C
2.	D
3.	D
4.	C
5.	B
6.	B
7.	D
8.	B
9.	C
10.	D
11.	E
12.	E
13.	C
14.	A
15.	E
16.	D
17.	C
18.	A
19.	C
20.	E
21.	A
22.	E
23.	B
24.	A
25.	B

UK SENIOR MATHEMATICAL CHALLENGE

Organised by the **United Kingdom Mathematics Trust**

SOLUTIONS

Keep these solutions secure until after the test on

TUESDAY 9 NOVEMBER 1999

This solutions pamphlet outlines a solution for each problem on this year's paper. We have tried to give the most straightforward approach, but the solutions presented here are not the only possible solutions. Occasionally we have added a 'Note' (in italics).

Please share these solutions with your students.

Much of the potential benefit of grappling with challenging mathematical problems depends on teachers making time for some kind of review, or follow-up, during which students may begin to see what they should have done, and how many problems they could have solved.

We hope that you and they agree that the first 15 problems could, in principle, have been solved by most candidates; if not, please let us know.

1. C The prime numbers less than 20 are 2, 3, 5, 7, 11, 13, 17, and 19.

(Prime numbers are defined as those numbers which have exactly two factors and therefore 1 is not prime since it has only one factor.)

2. D If January 1st (or, in a leap year, January 1st or January 2nd) falls on a Sunday, then there will be 53 Sundays in that particular year.

3. D The two shaded triangles will overlap when the 'net' is folded.

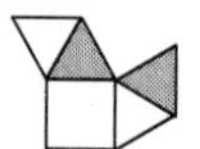

4. C The reductions would be A 20%; C $33\frac{1}{3}$ %; D 30%; E 25%. B might be a "special offer", but is not one which should be accepted!

5. B 17 hours and 21 minutes = 1041 minutes and the average speed is therefore slightly less than 40 decimal places per minute.

6. B The percentage increase $= \dfrac{500 - 40}{40} \times 100\% = \dfrac{460}{40} \times 100\% = 1150\%$.

7. D Let the exterior angle be $x°$. Then $x + 4x = 180$ and therefore $x = 36$.

As is the case in all convex polygons, the sum of the exterior angles = 360° and therefore the number of sides = 360/36 = 10.

8. B $x - y = 9$; $x + y = 99$. Adding gives $2x = 108$ and therefore $x = 54$ and hence $y = 45$.

The ratio $x{:}y = 54{:}45 = 6{:}5$.

(As we are interested only in the ratio rather than the numbers themselves, the problem could be reduced to finding the ratio of two numbers which differ by 1 and have sum 11.)

9. C All of the positive integers from 1 to 50 inclusive must be factors of 50!.

$51 = 3 \times 17$ and $52 = 2 \times 26$ which means that these two numbers are also factors of 50!.

53 is prime and is not a factor of 50!.

10. D The values are, approximately, A 45; B $1 \times 30 = 30$; C $20 \times 10 = 200$; D $200 \times 3 = 600$; E $2000 \times 0 = 0$.

11. E In the first row, any one of 5 letters could be circled. In the second row, any one of 4 letters could be circled since one column has now been occupied. Similarly, in the third row, any one of three letters could be circled and so on. The number of different ways is therefore $5 \times 4 \times 3 \times 2 \times 1 = 120$.

(Notice that this is 5! as in 9.)

12. E The White Rabbit must have been speaking on January 1st of this year. Two days earlier, December 30th, Alice was still thirteen and her fourteenth birthday was the following day, December 31st. She will, therefore, be fifteen on December 31st of this year and her sixteenth birthday will be on December 31st next year.

13. C As we are given, $AO = \frac{1}{3}AC$; therefore $AA' = \frac{2}{3}AC$ since $OA' = OA$.

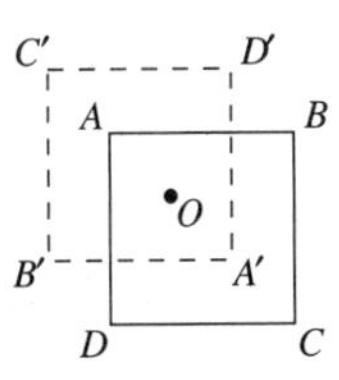

Thus the area of overlap is the area of a square whose side is $\frac{2}{3}$ of the length of the side of square $ABCD$ i.e. the area of a square of side 2 cm.

14. A The line $y = 3x + 4$ intersects the axes at $\left(-\frac{4}{3}, 0\right)$ and $(0, 4)$. Its reflection in the line $y = -x$ will therefore intersect the axes at $(-4, 0)$ and $(0, \frac{4}{3})$. The line through these points has gradient $\frac{1}{3}$ and therefore its equation is $y = \frac{1}{3}x + \frac{4}{3}$ or $3y = x + 4$.
(In general, the image of the point (a, b) after reflection in the line $y = -x$ is the point $(-b, -a)$.)

15. E $120 = 2^3 \times 3 \times 5$; $144 = 2^4 \times 3^2$; $240 = 2^4 \times 3 \times 5$; $3000 = 2^3 \times 3 \times 5^3$; $12100 = 2^2 \times 5^2 \times 11^2$. The product of the three numbers must have 6 prime factors, not necessarily all different, but with no prime factor repeated more than 3 times. Of these, only 12100 satisfies this condition. The three numbers are 10 (2×5), 22 (2×11) and 55 (5×11).

16. D x must be closer to 1000 than it is to 999 and also closer to 1000 than it is to 1010.

17. C Let Jon's and Jan's ages be $3x$ and x respectively. Then $3x - 3 = 4(x - 3)$ which gives $x = 9$. Therefore Jon is 27 and Jan is 9. If the ratio will be 2 : 1 in y years time then $27 + y = 2(9 + y)$ which gives $y = 9$. In 9 years time, Jon will be 36 and Jan will be 18.

18. A Let the radii of the outer and inner circles be R and r respectively. Then, by the Theorem of Pythagoras: $R^2 = r^2 + p^2$ and therefore $R^2 - r^2 = p^2$. The area of the shaded region $= \pi R^2 - \pi r^2 = \pi\left(R^2 - r^2\right) = \pi p^2$.

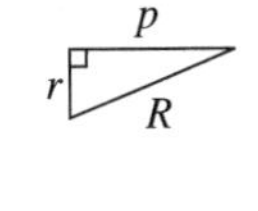

19. C By the Theorem of Pythagoras, $QR = \sqrt{8}\,\text{cm} = 2\sqrt{2}\,\text{cm}$, $QS = \sqrt{5}\,\text{cm}$ and $RS = \sqrt{5}$ cm. If T is the midpoint of QR, then $TS^2 = (\sqrt{5})^2 - (\sqrt{2})^2 = 5 - 2 = 3$ and therefore $TS = \sqrt{3}$ cm. The area of triangle $QRS = \frac{1}{2}QR \times TS = \left(\sqrt{2} \times \sqrt{3}\right)\text{cm}^2 = \sqrt{6}\,\text{cm}^2$.

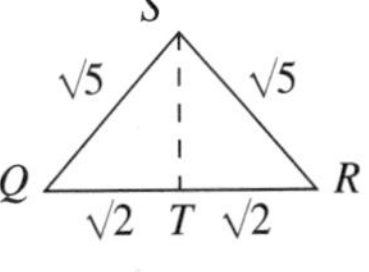

20. E $1 + 2 + 3 + 4 + 5 + \ldots + n = \frac{1}{2}n(n + 1)$.
Therefore the terms in the sequence with positions $\frac{1}{2}(n - 1)n + 1$ to $\frac{1}{2}n(n + 1)$ inclusive will all be n. Now $\frac{1}{2} \times 62 \times 63 = 1953$ and $\frac{1}{2} \times 63 \times 64 = 2016$. Therefore the 1954th term to the 2016th term inclusive will all be 63.

21. A B cannot be prime since 111 is clearly a factor of it; C cannot be prime since it equals 2999×1001 (difference of two squares); the units digit of D will be 5 and therefore 5 must be a factor of it whilst the units digit of E is 0 and therefore it cannot be prime either since both 2 and 5 will be factors of it.

22. E Let the radius of each large semicircle be R. Then $\pi R^2 = 4$. The circle in the left-hand diagram has radius $R/2$ and therefore its area is $\pi(R/2)^2 = (\pi R^2)/4 = 1$.

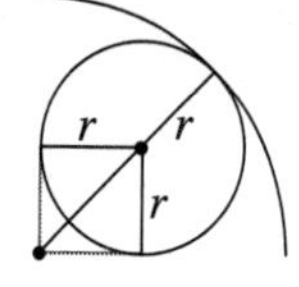

Let the radii of the circles in the right-hand diagram be r. Then $R = \sqrt{2}r + r$.

Therefore $r = \dfrac{R}{\sqrt{2}+1} = \dfrac{R(\sqrt{2}-1)}{(\sqrt{2}+1)(\sqrt{2}-1)} = R(\sqrt{2}-1)$.

The grey shaded area

$$= 2\pi r^2 = 2\pi R^2(\sqrt{2}-1)^2 = 2 \times 4 \times (2 - 2\sqrt{2} + 1) = 24 - 16\sqrt{2}.$$

The difference in areas is therefore $23 - 16\sqrt{2}$.

23. B
$$\frac{2n+y}{n-2} = \frac{2n-4}{n-2} + \frac{y+4}{n-2} = 2 + \frac{y+4}{n-2} \qquad (n \neq 2).$$

Thus $(2n + y)/(n - 2)$ is an integer if and only if $(y + 4)/(n - 2)$ is an integer. As n varies, the integer values taken by $(y + 4)/(n - 2)$ are all the integers which divide exactly into $y + 4$. There are exactly 4 of these if and only if $y + 4$ is prime. They are $\{(y + 4), -(y + 4), 1, -1\}$. Thus the integer values of the expression will be $2 + y + 4 = y + 6$; $2 - (y + 4) = -(y + 2)$; 3 and 1.

For $1 \leqslant y \leqslant 20$, the values of y for which $y + 4$ is prime are 1, 3, 7, 9, 13, 15 and 19.

24. A If the original tile is an isosceles trapezium made from three equilateral triangles then a quadrilateral (which is also an isosceles trapezium) will result.

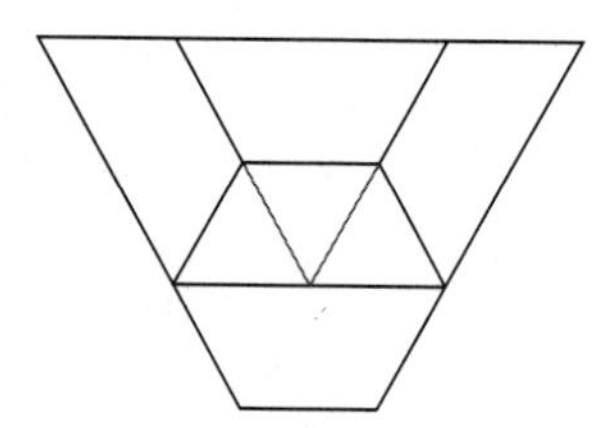

25. B
$$S = \frac{1}{2} + \frac{1}{4} + \frac{2}{8} + \frac{3}{16} + \frac{5}{32} + \frac{8}{64} + \frac{13}{128} + \frac{21}{256} + \frac{34}{512} + \ldots$$
$$= \frac{1}{2} + \frac{1}{4} + \left(\frac{1}{8} + \frac{1}{8}\right) + \left(\frac{1}{16} + \frac{2}{16}\right) + \left(\frac{2}{32} + \frac{3}{32}\right) + \left(\frac{3}{64} + \frac{5}{64}\right) + \ldots$$
$$= \frac{1}{2} + \left(\frac{1}{8} + \frac{1}{16} + \frac{2}{32} + \frac{3}{64} + \ldots\right) + \left(\frac{1}{4} + \frac{1}{8} + \frac{2}{16} + \frac{3}{32} + \frac{5}{64} + \ldots\right)$$
$$= \frac{1}{2} + \frac{1}{4}\left(\frac{1}{2} + \frac{1}{4} + \frac{2}{8} + \frac{3}{16} + \ldots\right) + \frac{1}{2}\left(\frac{1}{2} + \frac{1}{4} + \frac{2}{8} + \frac{3}{16} + \frac{5}{32} + \ldots\right)$$
$$= \frac{1}{2} + \frac{1}{4}S + \frac{1}{2}S = \frac{1}{2} + \frac{3}{4}S$$

Therefore $\quad \dfrac{1}{4}S = \dfrac{1}{2} \quad$ and so $\quad S = 2.$

As the papers are marked at the presenting centres it is not possible to comment on individual questions. Overall the paper was harder than in 1998 so the cut-offs were higher. On the basis of the standard proportions used by the UKMT, the cut-off marks were set at

GOLD – 80 or over SILVER – 65 to 79 BRONZE – 53 to 64

List of high scorers in the 1999 UK Senior Challenge

125	Andrew Smith	Ilford County HS
	David Collier	King Edward VI Southampton
	David Knipe	Sullivan Upper School
121	Stephen Brooks	Abingdon School
	Stuart Haring	Haberdashers Askes School
	Keita Chiba	Judd School
120	Peter Allen	Nottingham HS
	David Hodge	Torquay Boys' GS
	Kerwin Hui	Berkhamsted Collegiate
	Christopher Cummins	Our Lady of Sion School
117	Amelia Taylor	Wakefield Girls' HS
	Andrew McVitty	Sullivan Upper School
	JNR Williams	Winchester College
116	Do-Hyun Lee	Methodist College Belfast
	Liyang Au	Methodist College Belfast
	Bryn Garrod	King Edward VI Camp Hill Boys School
	Jacob Steel	Colchester RGS
	Ying Tao	KE VI Camp Hill Girls School
	T. Barnet-Lamb	Westminster School
115	Mark Woodward	George Ward School
	Alex Scordellis	RGS Guildford
	James Hilder	Durham Johnston CS
	Jack Bickeridge	RGS Worcester
	Matthew Chandler	Gravesend GS
	ORH Thomas	Winchester College
	Andrew Young	King Edward VI Camp Hill Boys School

114	Navindu Katugampola	Dulwich College
	Manuel Starr	Wilson's School
113	Ben Davis	Wells Cathedral School
	Fenton Whelan	Dulwich College
	Andrew Lim	Eton College
	Huan Gad	Loughborough GS
	Christopher Rayson	Netherhall School
	Tim West	Portsmouth GS
	Andrew Ross	Dr Challoner's GS
	Douglas Speed	Whitgift School
	DWA Wilson	Royal Belfast Academical Institute
112	Vincent Lee	Warwick School
	John Cassidy	Dulwich College
	Will W Macnair	High Storrs School
	Grace Chen	High Storrs School
	Niall Sayers	Bristol Grammar School
	David Rufino	Bradford Grammar School
	Thomas Poole	Barton Peveril School
	Paul Cooper	SEEVIC College
	Mark Datta	Kings' College School
	AR Brown	Winchester College

A sample of one of the certificates is shown below.

UK SENIOR MATHEMATICAL CHALLENGE
1999

. . . .
of

. . . .
received a

CERTIFICATE for BEST IN SCHOOL

Peter Neumann

Chairman United Kingdom Mathematics Trust

THE UNITED KINGDOM SENIOR MATHEMATICAL CHALLENGE

The UK SMC encourages mathematical reasoning, precision of thought, and fluency in using basic mathematical techniques to solve non-standard problems. It is targeted at sixteen to eighteen year olds with a genuine interest in mathematics.

The problems on the UK SMC are designed to make students think, and sometimes smile. Most are accessible to younger students, yet still challenge those with more experience; they are also meant to be memorable and enjoyable.

Mathematics controls more aspects of the modern world than most people realise — from CDs, cash machines, telecommunications and airline booking systems to production processes in engineering, efficient distribution and stock-holding, investment strategies and 'whispering' jet engines. The scientific and industrial revolutions flowed from the realisation that mathematics was both the language of nature, and also a way of analysing — and hence controlling — our environment. In the last fifty years old and new applications of mathematical ideas have transformed the way we live.

All these developments depend on mathematical thinking — a mode of thought whose essential style is far more permanent than the wave of technological change which it has made possible. The problems on the UK SMC reflect this style, which pervades all mathematics, by challenging students to think clearly about simple, yet unfamiliar problems.

The UK SMC was established as the National Mathematics Contest in 1961. In 1998 there were 31,000 participants from 982 schools and colleges. Certificates are awarded to the highest scoring 40% of candidates (6% Gold, 13% Silver, 21% Bronze).

The next stage

Candidates who obtained a score of 93 or over in the 1999 Senior Challenge, which, by coincidence was the same score as in 1998, were invited to take the British Mathematical Olympiad Round 1. Within the UKMT, the British Mathematical Olympiad Committee has control of the papers and everything pertaining to them. The BMOC produces an annual account of its events which, once again, was edited by Dr Ben Meisner (Oundle School). Much of this account is included in the following pages. The cover is shown below.

BRITISH MATHEMATICAL OLYMPIAD

Round 1 : Wednesday, 12 January 2000

Time allowed *Three and a half hours.*

Instructions

- *Full written solutions – not just answers – are required, with complete proofs of any assertions you may make. Marks awarded will depend on the clarity of your mathematical presentation. Work in rough first, and then draft your final version carefully before writing up your best attempt.*

  Do not hand in rough work.

- *One* complete *solution will gain far more credit than several unfinished attempts. It is more important to complete a small number of questions than to try all five problems.*

- *Each question carries 10 marks.*

- *The use of rulers and compasses is allowed, but calculators and protractors are forbidden.*

- *Start each question on a fresh sheet of paper. Write on one side of the paper only. On each sheet of working write the number of the question in the top* left *hand corner and your name, initials and school in the top* right *hand corner.*

- *Complete the cover sheet provided and attach it to the front of your script, followed by the questions 1, 2, 3, 4, 5 in order.*

- *Staple all the pages neatly together in the top* left *hand corner.*

Do not turn over until told to do so.

BRITISH MATHEMATICAL OLYMPIAD

Round 1 : Wednesday, 12 January 2000

1. Two intersecting circles C_1 and C_2 have a common tangent which touches C_1 at P and C_2 at Q. The two circles intersect at M and N, where N is nearer to PQ than M is. The line PN meets the circle C_2 again at R. Prove that MQ bisects angle PMR.

2. Show that, for every positive integer n,

$$121^n - 25^n + 1900^n - (-4)^n$$

is divisible by 2000.

3. Triangle ABC has a right angle at A. Among all points P on the perimeter of the triangle, find the position of P such that

$$AP + BP + CP$$

is minimised.

4. For each positive integer k, define the sequence $\{a_n\}$ by

$$a_0 = 1 \qquad \text{and} \qquad a_n = kn + (-1)^n a_{n-1} \qquad \text{for each } n \geqslant 1.$$

Determine all values of k for which 2000 is a term of the sequence.

5. The seven dwarfs decide to form four teams to compete in the Millennium Quiz. Of course, the sizes of the teams will not all be equal. For instance, one team might consist of Doc alone, one of Dopey alone, one of Sleepy, Happy and Grumpy as a trio, and one of Bashful and Sneezy as a pair. In how many ways can the four teams be made up? (The order of the teams or of the dwarfs within the teams does not matter, but each dwarf must be in exactly one of the teams.)

Suppose Snow White agreed to take part as well. In how many ways could the four teams then be formed?

The British Mathematical Olympiad 2000
Round 1

The first round of the British Mathematical Olympiad was held on 12th January 2000. The paper was marked by volunteers over a weekend at the end of January in Windsor, organised by Brian Wilson and Christine Farmer. Although difficult, the paper still allowed many candidates to display great skill and knowledge and they deserve much commendation for their efforts. Below is a list of the prize winners from Round 1.

Mark Adcock	Westminster School
Hazel Alcraft	Stockport Grammar School
Peter Allen	Nottingham High School
Chris Arcoumanis	St Paul's School
Naji Ashry	City of London School
Timothy Austin	Colchester RGS
Ross Axe	King Ed VI Camp Hill BS
Rahil Baber	King Ed VI S (Morpeth)
Robert Backhouse	King Ed VI Camp Hill BS
Thomas Barnet-Lamb	Westminster School
Nathan Bowler	Knutsford High School
Stephen Brooks	Abingdon School
Adam Brown	Winchester College
Felicity Bulmer	Cheltenham Ladies College
Stephen Burgess	Judd School
Hannah Burton	City of London Girls S
Joe Catling	Christleton High School
Edward Catmur	Hinchley Wood School
Dominique Chan	Chigwell School
Hang-Jin Chang	Westminster School
Suli Chen	Brooke House College
Ping Chen	St. Lawrence College
Keita Chiba	Judd School
James Chilcott	St Paul's School
Sunny Chiu-Webster	Trinity School
David Chow	Clifton College
Christine Chui	Harrogate Ladies' College
David Collier	King Ed VI School (So't'n)

Imran Coomaraswamy	Westminster School
Paul Cooper	SEEVIC
Simon Dambe	Sevenoaks School
Barnaby Dawson	Tiffin School
Michael Dore	Latymer Upper School
Chris Dowden	Hereford Cathedral School
Rosie Eade	Colchester Sixth Form Coll
Robert Edwards	Colchester RGS
Thomas Fenlon	Bishop Luffa School
Andrew Fisher	Eton College
John Fremlin	Colchester RGS
Tom Garnett	Hills Road Sixth Form Coll
Bryn Garrod	King Ed VI Camp Hill BS
Joe Gray	Newent Community School
Stuart Haring	Haberdashers' Aske's S
David Hodge	Torquay Boys' GS
Yang Yang Hou	Millfield School
Kerwin Hui	Berkhamsted Collegiate S
Carl James	Leicester Grammar School
Paul Jefferys	Berkhamsted Collegiate S
Wenbo Jia	Fettes College
Gwyn Jones	Brighton, Hove & Sussex SFC
David Knipe	Sullivan Upper School
Felix Lai	Dulwich College
Daniel Lamy	Nottingham High School
Jack Langdon	Warwick School
James Law	King Ed VI Camp Hill BS
Yuan Li	Atlantic College
Yunpeng Li	Atlantic College
Henry Lockwood	College of Richard Collyer
David Loeffler	Coltham School
Tristan Marshall	Tiffin School
Andrew Mills	Portora Royal
Wentao Mo	Atlantic College
Alexander Moore	St Olave's Grammar School
Mark Overton	Eton College

Victoria Pinnion	Skegness Grammar School
Thomas Poole	Barton Peveril College
Matthew Price	K. Edward's C, Stourbridge
Christopher Rayson	Netherhall S (Cumbria)
Miklos Reiter	Rugby School
Jeremy Reizenstein	St Paul's School
Christina Reynolds	Wimbledon High School
Michael Richards	Davenport HS for Boys
David Rufino	Bradford Grammar School
Amy Russell	Westminster School
Jeremy Sadler	University College School
Niall Sayers	Bristol Grammar School
Alexander Scordells	RGS, Guildford
Yew Juan See	Sevenoaks School
Edward Segal	Fortismere School
David Sher	St Paul's School
Andrew Smith	Ilford County High School
Edward Smith	Haberdashers' Aske's S
Mostafa Lameen Souag	Eton College
Douglas Speed	Whitgift School
Jonathan Storey	Nottingham High School
Elena Stoyanova	Colchester Sixth Form Coll
Atsushi Tateno	Cranleigh School
Oliver Thomas	Winchester College
Simon Thomas	Kingston Grammar School
Donald Tse	Winchester College
Frederick Van der Wyck	Westminster School
Jack Vickeridge	RGS, Worcester
Edward Wallace	Graveney School
Tim West	Portsmouth GS
Jack Willis	Churston Ferrers GS
Edward Wood	K. Edward's S Birmingham
Liam Wren-Lewis	Torquay Boys' GS
Chenyun Yin	Graveney School
Shuizi Yu	Greenhead College
Wenjing Zhong	Concord College

BRITISH MATHEMATICAL OLYMPIAD

Round 2 : Tuesday, 29 February 2000

Time allowed *Three and a half hours.*

Each question is worth 10 marks.

Instructions

- *Full written solutions – not just answers – are required, with complete proofs of any assertions you may make. Marks awarded will depend on the clarity of your mathematical presentation. Work in rough first, and then draft your final version carefully before writing up your best attempt.*

  Rough work should be handed in, but should be clearly marked.

- *One or two* complete *solutions will gain far more credit than partial attempts at all four problems.*

- *The use of rulers and compasses is allowed, but calculators and protractors are forbidden.*

- *Staple all the pages neatly together in the top* left *hand corner, with questions 1, 2, 3, 4 in order, and the cover sheet at the front.*

In early March, twenty students will be invited to attend the training session to be held at Trinity College, Cambridge (6-9 April). On the final morning of the training session, students sit a paper with just 3 Olympiad-style problems. The UK Team – six members plus one reserve – for this summer's International Mathematical Olympiad (to be held in South Korea, 13-24 July) will be chosen immediately thereafter. Those selected will be expected to participate in further correspondence work between April and July, and to attend a short residential session before leaving for South Korea.

Do not turn over until told to do so.

BRITISH MATHEMATICAL OLYMPIAD

Round 2 : Tuesday, 29 February 2000

1. Two intersecting circles C_1 and C_2 have a common tangent which touches C_1 at P and C_2 at Q. The two circles intersect at M and N, where N is nearer to PQ than M is. Prove that the triangles MNP and MNQ have equal areas.

2. Given that x, y, z are positive real numbers satisfying $xyz = 32$, find the minimum value of

$$x^2 + 4xy + 4y^2 + 2z^2.$$

3. Find positive integers a and b such that

$$\left(\sqrt[3]{a} + \sqrt[3]{b} - 1\right)^2 = 49 + 20\sqrt[3]{6}.$$

4. (a) Find a set A of ten positive integers such that no six distinct elements of A have a sum which is divisible by 6.

(b) Is it possible to find such a set if "ten" is replaced by "eleven"?

The British Mathematical Olympiad 2000

Round 2

The second round of the British Mathematical Olympiad was held on 29th February 2000. Some of the top scorers from this round were invited to a residential course at Trinity College, Cambridge.

Round 2 Leading Scores

36	Thomas Barnet-Lamb	Westminster School
32	Stephen Brooks	Abingdon School
32	David Collier	King Edward VI School (Southampton)
25	Kerwin Hui	Berkhamsted Collegiate School
24	Oliver Thomas	Winchester College
23	Sunny Chiu-Webster	Trinity School
23	Paul Cooper	SEEVIC
21	Paul Jefferys	Berkhamsted Collegiate School
20	Hannah Burton	City of London Girls School
18	Michael Spencer	Lawnswood High School

After the Trinity Weekend a squad of eight was selected for IMO training: Thomas Barnet-Lamb, Stephen Brooks, Hannah Burton, David Collier, Kerwin Hui, David Knipe, Michael Spencer, Oliver Thomas. The Team Leader for this year's IMO in Korea was Imre Leader and the Deputy Leader was Richard Atkins. A full account of the IMO appears later in this book.

INTRODUCTION TO THE PROBLEMS AND FULL SOLUTIONS

These solutions are the result of many hours of work by a large number of people. They have been subject to many drafts and revisions. As such, they do not resemble the first jottings, failed ideas and discarded pages of rough work with which any solution is started. Before looking at the solutions pupils and teachers are encouraged to make a good effort to solve the problems by themselves. Without wrestling with the problem oneself, it is hard to develop a feeling for the question, to understand where the difficulties lie and to appreciate why one method of attack is successful while another may fail. Reading these solutions without first attempting the question is unlikely to be of much benefit.

Many thanks are due to the contestants and to the members of the committee who contributed variety, refinement, inventiveness and precision to these written solutions.

BMO 1 Questions and Solutions

1. Two intersecting circles C_1 and C_2 have a common tangent which touches C_1 at P and C_2 at Q. The two circles intersect at M and N, where N is nearer to PQ than M is. The line PN meets the circle C_2 again at R. Prove that MQ bisects angle PMR.

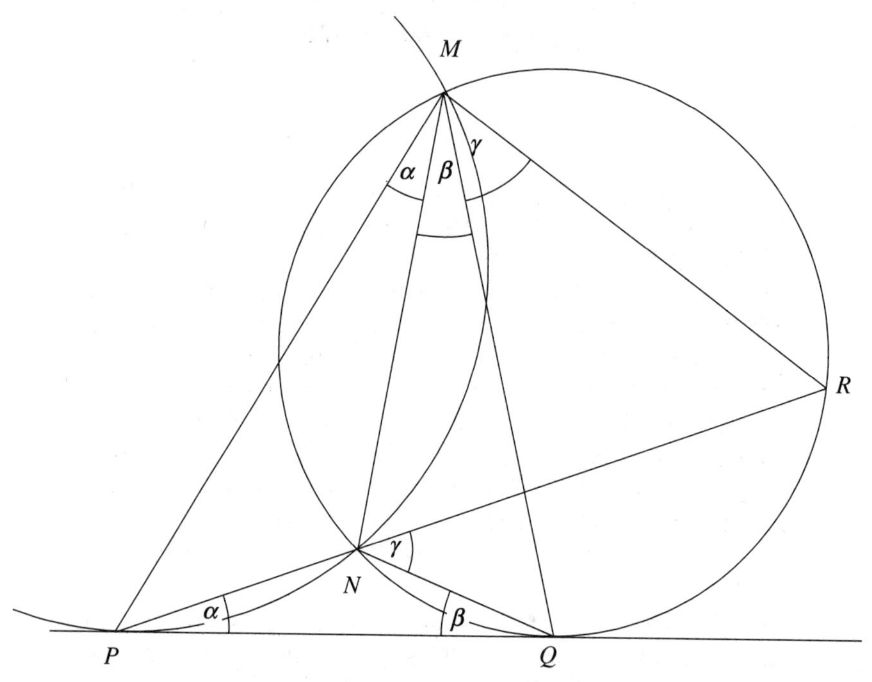

Join the points M and N.

Let $\angle PMN = \alpha$, $\angle NMQ = \beta$ and $\angle QMR = \gamma$.
Then $\angle QPN = \alpha$ and $\angle PQN = \beta$ by the alternate segment theorem applied respectively to the circles C_1 and C_2.
Furthermore $\angle QNR = \gamma$ since angles QMR and QNR are subtended by the same chord, QR, in the circle C_2.
The external angle, $\angle QNR$, of the triangle $\angle PNQ$, is equal to the sum of the internal opposite angles, $\angle QPN$ and $\angle PQN$.
Hence $\alpha + \beta = \gamma$ and $\angle PMQ = \angle QMR$ as required.

Alternative Solution (P. S. Cooper, SEEVIC)

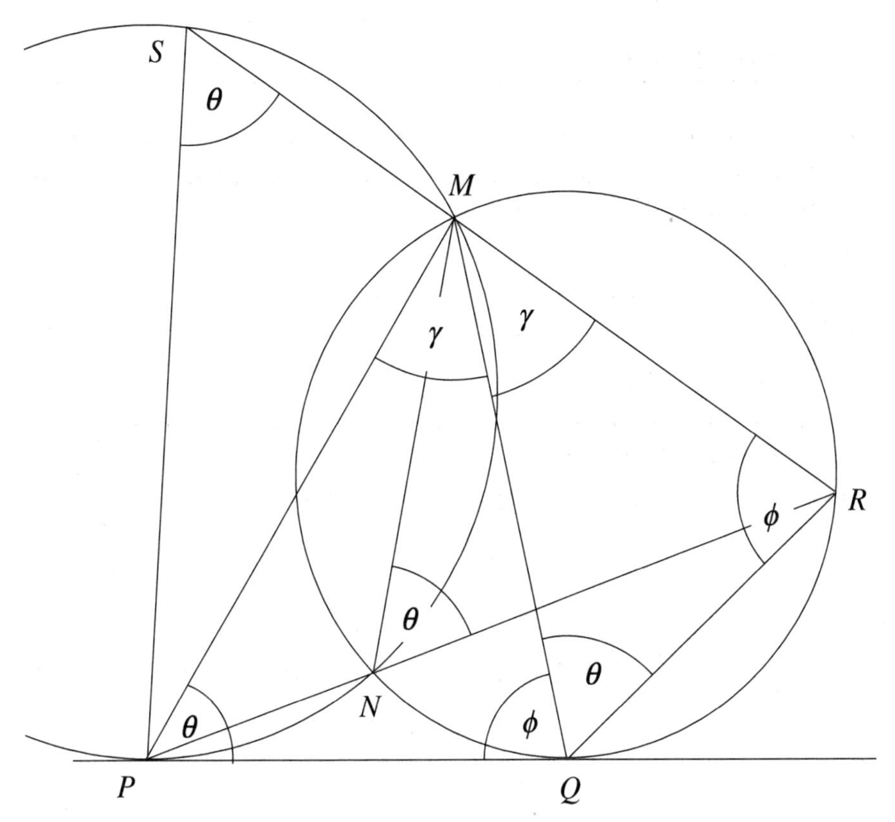

Produce the chord RM to meet C_1 at S. Join the lines MN, SP and RQ.

Let $\angle MPQ = \theta$, $\angle MQP = \phi$ and $\angle PMQ = \gamma$. Note that $\theta + \phi + \gamma = 180°$ (angle sum of triangle PQM).

By the alternate segment theorem $\angle PSM = \theta$, and $\angle QRM = \phi$.

Now $\angle RNM = \angle PSM = \theta$ (exterior angle of cyclic quadrilateral $SMNP$) and $\angle MQR = \angle MNR = \theta$ (angles subtended by the same chord, MR).

Hence $\angle QMR = 180° - \theta - \phi = \gamma$ (angle sum of triangle MQR).

We have now shown that $\angle QMR = \angle PMQ$ and the proof that MQ bisects $\angle PMR$ is complete.

2. Show that, for every positive integer n,

$$121^n - 25^n + 1900^n - (-4)^n$$

is divisible by 2000.

We first note that, since $2000 = 2^4 \times 5^3$ and the highest common factor of 2^4 and 5^3 is 1, it suffices to prove that the given expression is divisible by $16 = 2^4$ and $125 = 5^3$ separately.

Method 1 (using factorisation)
We also note that any expression of the form $(x^n - y^n)$, where n is a positive integer, has $(x - y)$ as a factor:

$$(x^n - y^n) = (x - y)(x^{n-1} + x^{n-2}y + x^{n-3}y^2 + \dots + x^{n-r-1}y^r + \dots + y^{n-1}).$$

By pairing off terms in the given expression, we can deduce quickly that it is divisible by 16 and 125:

$$\begin{aligned} 121^n - 25^n + 1900^n - (-4)^n &= (121^n - 25^n) + (1900^n - (-4)^n) \\ &= 96c + 1904d, \text{ where } c \text{ and } d \text{ are integers} \\ &= 16(6c + 119d) \end{aligned}$$

$$\begin{aligned} 121^n - 25^n + 1900^n - (-4)^n &= (121^n - (-4)^n) + (1900^n - 25^n) \\ &= 125a + 1875b, \text{ where } a \text{ and } b \text{ are integers} \\ &= 125(a + 15b). \end{aligned}$$

Method 2 (using modular arithmetic)

$$121 \equiv 25 \pmod{16} \Rightarrow 121^n \equiv 25^n \pmod{16}$$

$$1900 \equiv -4 \pmod{16} \Rightarrow 1900^n \equiv (-4)^n \pmod{16}.$$

$$\text{Hence } 121^n - 25^n + 1900^n - (-4)^n \equiv 0 \pmod{16}.$$

$$\text{Similarly } 121 \equiv -4 \pmod{125} \Rightarrow 121^n \equiv (-4)^n \pmod{125}$$

$$1900 \equiv 25 \pmod{125} \Rightarrow 1900^n \equiv 25^n \pmod{125}.$$

$$\text{Hence } 121^n - 25^n + 1900^n - (-4)^n \equiv 0 \pmod{125}.$$

In each case we have shown that $121^n - 25^n + 1900^n - (-4)^n$ is divisible by 16 and 125 and hence that it is divisible by 2000.

3. Triangle ABC has a right-angle at A. Among all points P on the perimeter of the triangle, find the position of P such that $AP + BP + CP$ is minimised.

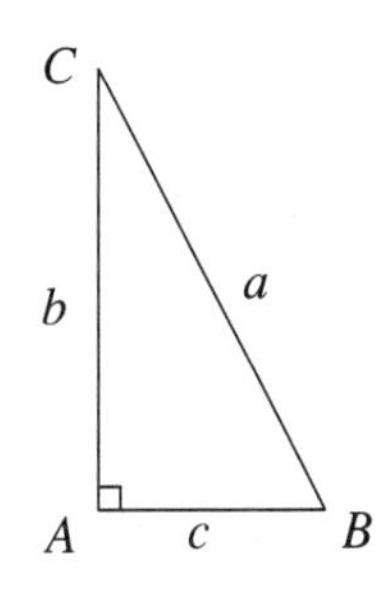

We start by labelling the lengths BC, AC, AB by a, b, c respectively as shown on the diagram.

We consider separately the possibilities: P is on AC, P is on AB, P is on BC.

If P is on AC, then $AP + CP = b$ so minimising $AP + BP + CP$ amounts to minimising BP. This occurs when P is at A, since the perpendicular distance of a point from a line is the least. The minimum of $AP + BP + CP$ in this case is $b + c$.

Similarly if P is on AB, then $AP + BP = c$ so that minimising $AP + BP + CP$ amounts to minimising CP. For the same reason, this also occurs when P is at A and the minimum of $AP + BP + CP$ is once again $b + c$.

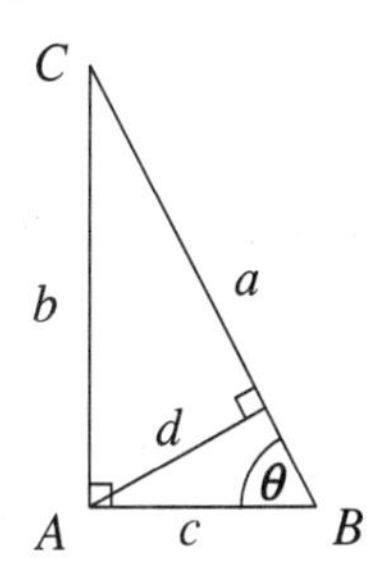

Now suppose that P is on BC. Then $BP + CP = a$ so that minimising $AP + BP + CP$ amounts to minimising AP. This happens when AP is perpendicular to BC. This length is labelled d in the diagram on the left. In this case the minimum of $AP + BP + CP$ is $a + d$.

The question now boils down to a comparison of $a + d$ with $b + c$.

Method 1

We note first that we can find the area of the triangle ABC in two ways, yielding $\frac{1}{2}bc = \frac{1}{2}ad$, which gives us $2bc = 2ad$.

By Pythagoras' Theorem $a^2 = b^2 + c^2$, and, since $d^2 > 0$,

we have $a^2 + d^2 > b^2 + c^2$.

Hence $\quad a^2 + 2ad + d^2 > b^2 + 2bc + c^2$

i.e. $\quad (a + d)^2 > (b + c)^2$

i.e. $\quad a + d > b + c$ since all quantities are non-negative.

Method 2

Let angle $ABC = \theta$. Then $d = c \sin\theta$ and $b = a \sin\theta$. Since BC is the hypotenuse, $a > c$ and, since $0° < \theta < 90°$, $1 - \sin\theta > 0$. Hence

$$a(1 - \sin\theta) > c(1 - \sin\theta)$$

i.e.
$$a + c\sin\theta > c + a\sin\theta$$

i.e.
$$a + d > b + c.$$

Method 3 (S. E. Thomas, Kingston Grammar School)

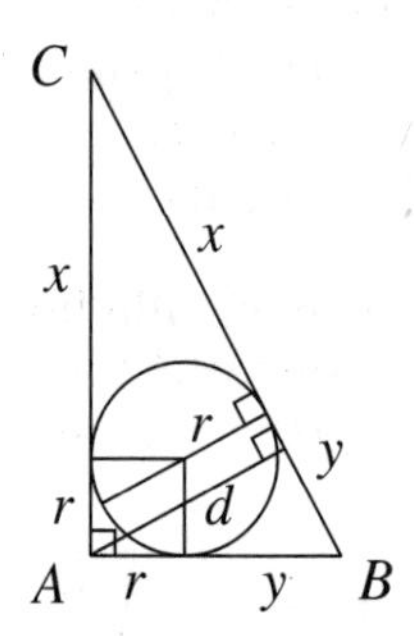

Consider the inscribed circle of triangle ABC.

Note that the distances from a point outside a circle along the two tangents to where they meet the circle are equal. We label the distances from points B and C along the sides of the triangle to the circle x and y respectively. We also label the radius of the circle r. This is the distance from the point A to the circle.

Then $b + c = x + y + 2r$ and $a + d = x + y + d$.
It is clear, for example by considering similar triangles, that $d > 2r$ and hence that $a + d > b + c$.

Conclusion

We conclude that the minimum of $AP + BP + CP$ is $b + c$ and occurs when P is at A.

4. For each positive integer k, define the sequence $\{a_n\}$ by

$$a_0 = 1, \qquad a_n = kn + (-1)^n a_{n-1} \quad \text{for each } n \geqslant 1.$$

Determine all values of k for which 2000 is a term of the sequence.

We prove that the terms in the sequence take the form

$$\begin{aligned} a_{4m} &= 4mk + 1 \\ a_{4m+1} &= k - 1 \\ a_{4m+2} &= (4m+3)k - 1 \\ a_{4m+3} &= 1 \qquad\qquad (m > 0). \end{aligned}$$

We start by considering the odd terms in the sequence.
If n is odd then, from the definition of the sequence,

$$\begin{aligned} a_{n+2} &= k(n+2) + (-1)^{n+2} a_{n+1} \\ &= k(n+2) - a_{n+1} \\ &= k(n+2) - \left(k(n+1) + (-1)^{n+1} a_n\right) \\ &= k(n+2) - k(n+1) - a_n \\ &= k - a_n \end{aligned}$$

Applying this formula twice, we have $a_{n+4} = k - a_{n+2} = k - (k - a_n) = a_n$.
Since $a_1 = k - 1$, the odd terms in the sequence are given by $a_{4m+1} = k - 1, a_{4m+3} = 1$.
Having established this relationship, the formulae for a_{4m+2} $(m \geqslant 0)$ and a_{4m} $(m \geqslant 0)$ follow from the definition of the sequence:

$$\begin{aligned} a_{4m+2} &= k(4m+2) + (-1)^{4m+2}(k-1) \\ &= k(4m+2) + (k-1) \\ &= (4m+3)k - 1 \end{aligned} \qquad \begin{aligned} a_{4m} &= k(4m) + (-1)^{4m} a_{4(m-1)+3} \\ &= 4mk + 1 \end{aligned}$$

We conclude that 2000 appears in the sequence if

- $2000 = 4mk + 1$ or $k = \dfrac{1999}{4m}$. This does not yield any integer solutions for k since 4 is not a factor of 1999.
- $2000 = k - 1$. This yields just one solution, $k = 2001$.
- $2000 = (4m+3)k - 1$ or $k = \dfrac{2001}{4m+3}$. The prime factors of 2001 are $2001 = 3 \times 23 \times 29$. The factors of 2001 are 1, 3, 23, 29, $3 \times 23 = 69$, $3 \times 29 = 87$, $23 \times 29 = 667$ and 2001. Of these only 3, 23, 87 and 667 have the form $4m + 3$, where m is an integer. These give 667, 87, 23, 3 as the only possible values of k in this case.
- $2000 = 1$. This gives no solutions.

The values of k for which 2000 is a term of the sequence are 3, 23, 87, 667, 2001.

5. The seven dwarfs decide to form four teams to compete in the Millennium Quiz. Of course, the sizes of the teams will not all be equal. For instance, one team might consist of Doc alone, one of Dopey alone, one of Sleepy, Happy and Grumpy as a trio, and one of Bashful and Sneezy as a pair. In how many ways can the four teams be made up?

   (The order of the teams or of the dwarfs within the teams does not matter, but each dwarf must be in exactly one of the teams.)

   Suppose Snow White agreed to take part as well. In how many ways could the teams then be formed?

We assume throughout that a team has at least one member and that all the players are assigned to a team.

Method 1 – Recurrence Solution

Let $n_{d,t}$ be the number of ways of forming t teams from d players, with $t \leqslant d$. In this notation, the question asks us to find $n_{7,4}$ and $n_{8,4}$.

There is only ever one way to form 1 team from d players, comprising all the players, so that $n_{d,1} = 1$.

There is only ever one way to form d teams from d players, where each player constitutes a complete team, so that $n_{d,d} = 1$.

Let us now single out a player, Bashful say, when we want to form t teams with d players (including Bashful), $1 < t < d$. There are 2 cases.

Bashful could be a team on his own. The number of ways of arranging this is $n_{d-1,t-1}$, the number of ways to form the remaining $t - 1$ teams from the other $d - 1$ players.

Alternatively Bashful could be in a larger team. We start by considering the other players. They can be placed into t teams, in $n_{d-1,t}$ ways. Now Bashful is added to one of these t teams. Each arrangement of the other $d - 1$ players into t teams and each addition of Bashful to one of these teams gives rise to a different arrangement of t teams with d players, yielding a total of $t \times n_{d-1,t}$ ways in which Bashful can be in a team with other players.

	d =	1	2	3	4	5	6	7	8
t = 1		1	1	1	1	1	1	1	1
t = 2			1	3	7	15	31	63	127
t = 3				1	6	25	90	301	966
t = 4					1	10	65	<u>350</u>	<u>1701</u>

Hence we have the recurrence relationship $n_{d,t} = n_{d-1,t-1} + t \times n_{d-1,t}$ which is used to calculate the values in the table above

We get the answers $n_{7,4} = 350$ and $n_{8,4} = 1701$.

Method 2 – Counting Solution

This solution relies first upon finding out the possible sizes of the teams and then counting up the number of ways of placing the dwarfs into teams of these sizes. It is essential that a method of counting is chosen so that each possible arrangement of teams is counted exactly once.

7 Dwarfs

There must always be at least one team of size 1 since $4 \times 2 = 8$ and there can be at most three teams of size 1. The largest possible team size is 4. Listing possible team sizes in descending order we get

4, 1, 1, 1 which can be done in ${}^7C_4 = 35$ ways. (Choose the team of 4 first, then the singles are made up from the remaining players.)

3, 2, 1, 1 which can be done in ${}^7C_3 \times {}^4C_2 = 210$ ways. (Choose the team of 3 first, then the team of 2 from the remaining 4, then the singles are made up from the remaining players.)

2, 2, 2, 1 which can be done in $\dfrac{{}^7C_2 \times {}^5C_2 \times {}^3C_2}{3!} = 105$ ways. (Choose the first team of 2 from 7, the second from 5, the third from 3, leaving the final team of 1. However the three teams of 2 could have been picked in any order which accounts for the division by 3! .)

This gives a total of 350 ways.

7 Dwarfs and Snow White

There can be at most three teams of size 1. The largest possible team size is 5. Listing possible team sizes in descending order we get

5, 1, 1, 1 which can be done in ${}^8C_5 = 56$ ways

4, 2, 1, 1 which can be done in ${}^8C_4 \times {}^4C_2 = 420$ ways

3, 3, 1, 1 which can be done in $\dfrac{{}^8C_3 \times {}^5C_3}{2!} = 280$ ways

3, 2, 2, 1 which can be done in ${}^8C_3 \times \dfrac{{}^5C_2 \times {}^3C_2}{2!} = 840$ ways

2, 2, 2, 2 which can be done in $\dfrac{{}^8C_2 \times {}^6C_2 \times {}^4C_2 \times {}^2C_2}{4!} = 105$ ways.

This gives a total of 1701 ways.

BMO 2 Questions and Solutions

1. Two intersecting circles C_1 and C_2 have a common tangent which touches C_1 at P and C_2 at Q. The two circles intersect at M and N, where N is nearer to PQ than M is. Prove that the triangles MNP and MNQ have equal areas.

Extend the line MN and label the point X at which it meets PQ. We set out to prove that $PX = XQ$.

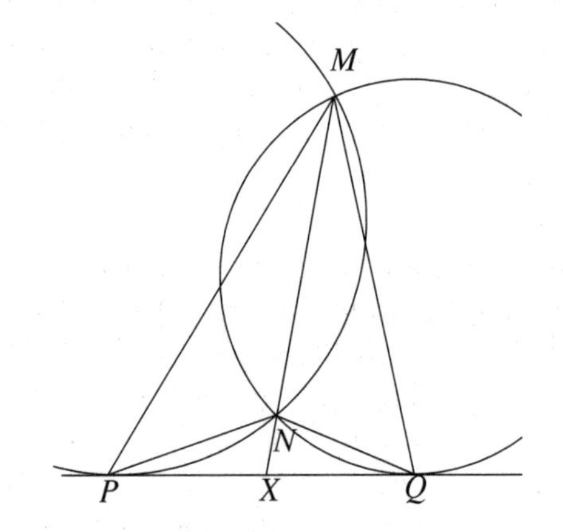

By the alternate segment theorem, we have that $\angle XPN = \angle PMN$. The angle MXP is common to triangles MXP and PXN. Hence these triangles are similar since they have all three angles equal. Since the ratio of lengths of corresponding sides of these triangles are equal, we have $\dfrac{XP}{XM} = \dfrac{XN}{XP}$ and $XP^2 = XN \times XM$.

By an identical argument, triangles MXQ and QMN are similar. Hence $\dfrac{XQ}{XM} = \dfrac{XN}{XQ}$ and $XQ^2 = XN \times XM$.

(N. B. the two results $XP^2 = XN \times XM$ and $XQ^2 = XN \times XM$ are each a special case of the intersecting chords theorem for circles and could be quoted as such.)

Thus PX and QX are both equal to $\sqrt{XN \times XM}$. It follows that the pair of triangles, MXP and MXQ, have the same area since their bases, PQ and QX, are equal and their heights, the perpendicular distance from PQ to M, are identical. Similarly the pair of triangles, NXP and NXQ have the same base and height. Since

$$\text{Area of } MNP = \text{Area of } MXP - \text{Area of } NXP,$$
$$\text{and Area of } MNQ = \text{Area of } MXQ - \text{Area of } NXQ,$$

the areas of triangles MNP and MNQ are equal as required.

2. Given that x, y, z are positive real numbers satisfying $xyz = 32$, find the minimum value of

$$x^2 + 4xy + 4y^2 + 2z^2.$$

The *Arithmetic Mean - Geometric Mean (AM-GM) Inequality* states that for any n non-negative real numbers, $x_1, x_2, \ldots, x_n$,

$$\frac{x_1 + x_2 + \ldots + x_n}{n} \geqslant (x_1 x_2 \ldots x_n)^{1/n},$$

with equality only when $x_1 = x_2 = \ldots = x_n$.

We can use this inequality in this question if we can somehow arrange for the right-hand side of the inequality to be some power of xyz. This is achieved by dividing up terms in the sum $x^2 + 4xy + 4y^2 + 2z^2$. One such division is $x^2 + 2xy + 2xy + 4y^2 + z^2 + z^2$ since these terms multiply to give $16\,(xyz)^4$.

Now

$$\begin{aligned} x^2 + 4xy + 4y^2 + 2z^2 &= 6 \times \frac{x^2 + 2xy + 2xy + 4y^2 + z^2 + z^2}{6} \\ &\geqslant 6 \times \left(16(xyz)^4\right)^{1/6}, \text{ by the AM-GM inequality.} \\ &= 6 \times \left(16 \times 32^4\right)^{1/6} \\ &= 6 \times \left(2^{24}\right)^{1/6}, \text{ since } 16 = 2^4 \text{ and } 32 = 2^5 \\ &= 6 \times 2^4 = 96. \end{aligned}$$

This shows that the minimum possible value is 96. This minimum is achieved only if all the terms in the sum $x^2 + 2xy + 2xy + 4y^2 + z^2 + z^2$ are equal, i.e. $x^2 = 2xy = 4y^2 = z^2$. This does occur with $xyz = 32$ when $x = z = 4$ and $y = 2$, so we conclude that the minimum is 96.

Alternative Solution (Edward Wallace, Graveney School)

This method uses inventive ways to reduce the numbers of variables under consideration, first by using the constraint $xyz = 32$ and then by noticing a property of the inequality found.

We first rearrange the expression to be minimised;

$$x^2 + 4xy + 4y^2 + 2z^2 = (x + 2y)^2 + 2z^2.$$

Let $u = (x + 2y)^2$ and $v = (x - 2y)^2$.

Then $u - v = (x + 2y)^2 - (x - 2y)^2 = 8xy$, so that $z = \dfrac{32}{xy} = \dfrac{256}{u - v}$.

Hence $x^2 + 4xy + 4y^2 + 2z^2 = (x + 2y)^2 + 2z^2 = u + 2\left(\dfrac{256}{u - v}\right)^2 = u + \dfrac{2^{17}}{(u - v)^2}$.

Since u and v are independent variables and $v = (x - 2y)^2 \geqslant 0$, $u + \dfrac{2^{17}}{(u - v)^2}$ takes its minimum value when $v = 0$, if that is possible.

We differentiate $u + \dfrac{2^{17}}{u^2}$ to find its minimum. At the minimum point

$$\frac{d}{du}\left(u + \frac{2^{17}}{u^2}\right) = 0$$

$$1 - \frac{2^{18}}{u^3} = 0$$

$$u = \sqrt[3]{2^{18}} = 2^6 = 64.$$

This value of u gives

$$x^2 + 4xy + 4y^2 + 2z^2 = u + \frac{2^{17}}{u^2} = 64 + \frac{2^{17}}{2^{12}} = 64 + 32 = 96.$$

To confirm that this is a minimum value, we consider the value of the second derivative when $u = 64$. Since $\dfrac{3 \times 2^{18}}{u^4} > 0$, this is a local minimum. We need now to verify that this is a global minimum for the range of values of u under consideration. However, we note that the second derivative is positive for all u. Therefore $u + \dfrac{2^{17}}{u^2}$ is a convex function and hence this is a global maximum.

Now $(x + 2y)^2 = 64$, $(x - 2y)^2 = 0$ and $xyz = 32$ has the solution $x = z = 4, y = 2$.

Hence we conclude that 96 is the minimum possible value of $x^2 + 4xy + 4y^2 + 2z^2$.

3. Find positive integers a and b such that

$$\left(\sqrt[3]{a} + \sqrt[3]{b} - 1\right)^2 = 49 + 20\sqrt[3]{6}.$$

Solution 1 (Thomas Barnet-Lamb, Westminster School)
Let $a = 6x^3$ and $b = 36y^3$. Then $\sqrt[3]{a} = x\sqrt[3]{6}$ and $\sqrt[3]{b} = y\sqrt[3]{36}$. Noting that $\sqrt[3]{36^2} = 6\sqrt[3]{6}$, we get

$$\begin{aligned}\left(\sqrt[3]{a} + \sqrt[3]{b} - 1\right)^2 - 1 &= \left(x\sqrt[3]{6} + y\sqrt[3]{36} - 1\right)^2 - 1\\ &= \left(6y^2 - 2x\right)\sqrt[3]{6} + \left(x^2 - 2y\right)\sqrt[3]{36} + 12xy\\ &= 48 + 20\sqrt[3]{6}.\end{aligned}$$

Equating coefficients of $\sqrt[3]{6}$, $\sqrt[3]{36}$ and 1, we get the three equations $6y^2 - 2x = 20$, $x^2 - 2y = 0$ and $xy = 4$. These three equations are satisfied by the integer solution $x = y = 2$ which yields $a = 48$ and $b = 288$.

Solution 2
Expanding the left-hand side of the equation and subtracting 1 from both sides gives

$$\sqrt[3]{a^2} + \sqrt[3]{b^2} - 2\sqrt[3]{a} - 2\sqrt[3]{b} + 2\sqrt[3]{ab} = 48 + 20\sqrt[3]{6}.$$

Assuming that a and b are not cubes, the only term on the left which could be rational is $2\sqrt[3]{ab}$, so we will equate this with 48, giving $ab = 24^3 = 2^9 \times 3^3 = 2^6 \times 6^3$.

The remaining terms $\sqrt[3]{a^2} + \sqrt[3]{b^2} - 2\sqrt[3]{a} - 2\sqrt[3]{b}$ cancel to the relatively simple $20\sqrt[3]{6}$.
Assuming that two of the terms are equal, we try $\sqrt[3]{a^2} = 2\sqrt[3]{b}$, i.e. $a^2 = 8b$. This, together with $ab = 2^6 \times 6^3$, gives $a = 2^3 \times 6$ and $b = 2^3 \times 6^2$. With these values of a and b

$$\begin{aligned}\sqrt[3]{b^2} - 2\sqrt[3]{a} &= \sqrt[3]{2^6 \times 6^4} - 2\sqrt[3]{2^3 \times 6}\\ &= (4 \times 6)\sqrt[3]{6} - (2 \times 2)\sqrt[3]{6}\\ &= 20\sqrt[3]{6}\end{aligned}$$

as required.
Thus $a = 48$ and $b = 288$ are positive integers such that $\left(\sqrt[3]{a} + \sqrt[3]{b} - 1\right)^2 = 49 + 20\sqrt[3]{6}$.

4. (a) Find a set A of ten positive integers such that no six distinct elements of A have a sum which is divisible by 6.
   (b) Is it possible to find such a set if "ten" is replaced by "eleven"?

(a) The simplest example of such a set is $\{0,0,0,0,0,1,1,1,1,1\}$ if repetitions and zero are allowed. It is clear that any set of six of these numbers adds up to something greater than 0, since at least one 1 must be included, and less than 6, since at most five 1s can be included. Any set of numbers, five with remainder 0 when divided by 6 and five with remainder 1 when divided by 6 will do, for the same reason, e.g. $\{6,12,18,24,30,1,7,13,19,25\}$.

(b) It is not possible to find a set of 11 positive integers such that no six distinct elements have a sum divisible by six. To prove this, we prove two simpler, related results:

Lemma 1 Any set of 3 or more integers contains a pair whose sum is divisible by 2.

Proof Either the set contains at least two odd numbers or it contains at least two even numbers. The sum of such a pair is even.

Lemma 2 Any set of 5 or more integers contains a triple whose sum is divisible by 3.

Proof There are three different remainders when a number is divided by 3, namely 0, 1 or 2. If the set contains three numbers whose remainders when divided by 3 are all different, say $3a$, $3b + 1$, $3c + 2$, then the sum of these numbers, $3(a + b + c) + 3$, is divisible by 3. If not, then there are only two remainders possible for the five numbers. In this case at least three numbers must have the same remainder, say $3d + t$, $3e + t$, $3f + t$. The sum of these numbers, $3(d + e + f) + 3t$, is divisible by 3.

If we have eleven integers, then, by repeated application of Lemma 1, we can find five separate pairs of them, each with even sum. Applying Lemma 2 to these five integers, we can find three of these pairs whose sum is divisible by 3. This will also be even since it is the sum of even values. We now have six of the original integers whose sum is divisible by 6.

Olympiad Training Session
Trinity College, Cambridge 6-9 April 2000

On the basis of BMO2 and BMO1, 20 students are invited to take part in the Olympiad Training Session. This is a four-day event, held in Trinity College, Cambridge, and it serves a number of purposes.

First of all, it provides a large amount of mathematical stimulation, given in an intense burst. For three days the students have various sessions. Each day there are about three 90-minute courses, with a small amount of explanation at the beginnning but with the emphasis very much on the students tackling problems themselves!

Secondly, there is the social aspect. For many of the students, this is the first time they have been around other students who are as good, or perhaps even better, at mathematics than they are. The social mixing that goes on is always interesting to watch: although the group are mostly strangers at the start, they share a common interest in mathematics, so that very soon the room is buzzing with conversation.

Thirdly, the Training Session is part of the selection process for the United Kingdom team for the International Mathematical Olympiad. On the last day of the meeting the students sit a 4½ hour exam (not the way everyone would choose to spend a Sunday morning), consisting of three difficult questions. Based largely on this exam (the Final Selection Test), a squad of eight is chosen. The squad then embark on a correspondence course that will end up reducing them to a team of six.

The makeup of the 20 varies from year to year, but typically just over a half will be realistic contenders for this year's IMO team. The rest will be 'hopes for the future', not yet in their last year at school, who are there to soak up the atmosphere and the mathematics but perhaps also to challenge for a place in the squad.

The students assembled on the afternoon of Thursday 6th April, and things started off with an 'icebreaker' from Tony Gardiner. This is a puzzle designed to get people to relax and also meet their teammates. For the rest of Thursday, and also Friday, the students are split up into teams of four – each team will have a 'captain', who is usually an experienced hand.

So now the hard work starts. In each session, the students work as teams. Towards the end of day, some 'background problems' are given out. These are some fiendish problems by David Monk – each team has one problem, to think about right through to Sunday afternoon.

Friday goes as Thursday, with plenty of sessions and the students working in teams. That evening, as some light relief, a ‘relay’ is held. Each team has to race against the clock to solve some brainteasers – some running around the room is also involved, and there is general chaos! There are often one or two teams made up of ‘old Olympians’ – former IMO team members who have returned to offer advice amd encouragement. Nearly always it is one of those ‘extra’ old Olympiad teams that wins the relay, but this year one of the genuine teams beat all the oldies!

The schedule for Saturday is rather different. There are much smaller sessions, usually three in parallel. So about six or seven students attend each session, where they work individually. These sessions often develop themes touched on in the previous days' sessions.

After an early night, the next morning brings the FST. After lunch, the teams present their solutions or attempted solutions to the background problems, and there is then time for a farewell tea before departure. The staff tend to stay on for another 3 or 4 hours to mark the FST, so that by the end of the day the composition of the squad is well on the way to being decided.

The mathematics in the sessions covers several areas. These include algebra, number theory, geometry, combinatorics, inequalities and functional equations. The aim is always to keep the lecturing to a minimum, so that the time spent actually attacking problems is as great as possible! The sessions themselves are very informal: many of the staff will be wandering around the room, seeing how students are doing and chatting to them. Nearly everyone finds that the material they are shown at Trinity will keep them thinking for several weeks after the Training Session.

We are extremely grateful to Trinity College, who sponsor this event and make us feel so welcome, and particularly to Dr. Hugh Osborn. We are also very grateful to Julia Gog, who as well as being one of the staff does a huge amount of liaising with Trinity and general troubleshooting.

Outline Programme

Thursday 6 April

13.00-14.00	Arrival at Burrell's Field
14.00-14.30	Ice-Breaker (Tony Gardiner)
14.30-16.00	Introductory Problems (Imre Leader)
16.30-18.15	Algebra (Adam McBride)
18.30-19.15	*Dinner*
19.30-21.15	Number Theory 1 (Kevin Buzzard); Background Problems distributed (David Monk)

Friday 7 April

09.00-10.45	Combinatorics 1 (Imre Leader)
11.15-12.45	Geometry 1 (Richard Atkins)
13.00-14.00	*Lunch*
14.00-15.00	Trinity Talk and Prizes
16.30-18.15	Inequalities (Ben Green)
18.30-19.15	*Dinner*
19.30-21.30	Relays (Tony Gardiner)

Saturday 8 April

09.00-10.45	Functional Equations (Dan Crisan) Geometry 2 (David Monk) Combinatorics 2 (Imre Leader)
11.15-12.45	Geometric Inequalities (Christopher Bradley) Combinatorics 2 (Imre Leader) Number Theory 2 (Kevin Buzzard)
13.00-14.00	*Lunch*
16.30-18.15	Geometry 2 (David Monk) Number Theory 2 (Kevin Buzzard) Problem Solving (Tony Gardiner)
18.30-19.15	*Dinner followed by College Reception*

Sunday 9 April

08.15-12.45	Final Selection Test
13.00-14.00	*Lunch*
14.00-15.30	Background Problems discussed (David Monk)
15.30-16.00	Tea and departure from Burrell's Field

Participants

Robert Backhouse	King Edward VI Camp Hill School
Thomas Barnet-Lamb	Westminster School
Stephen Brooks	Abingdon School
Hannah Burton	City of London Girls School
Edward Catmur	Hinchley Wood School
Hang-Jin Chang	Westminster School
Sunny Chiu-Webster	Trinity School
David Collier	King Edward VI School, Southampton
Paul Cooper	SEEVIC
Ian Gallagher	Rugby School
Kerwin Hui	Berkhamsted Collegiate School
Paul Jefferys	Berkhamsted Collegiate School
David Knipe	Sullivan Upper School
Sue Liu	Madras College
David Loeffler	Cotham School
Victoria Pinnion	Skegness Grammar School
Edward Segal	Fortismere School
Michael Spencer	Lawnswood High School
Oliver Thomas	Winchester College
Edward Wallace	Graveney School

Staff

Richard Atkins	(Oundle School)
Christopher Bradley	(Clifton College)
Kevin Buzzard	(Imperial College, London)
Dan Crisan	(Queens' College, Cambridge)
Tony Gardiner	(University of Birmingham)
Julia Gog	(Trinity College, Cambridge)
Ben Green	(Trinity College, Cambridge)
Imre Leader	(University College London)
Adam McBride	(University of Strathclyde)
David Monk	(University of Edinburgh)

Domestic Arrangements:

Julia Gog (Trinity College, Cambridge)

41st International Mathematical Olympiad
Taejon, South Korea
13-25 July 2000

Report by Imre Leader (UK Team Leader)

This is a report on the 41st International Mathematical Olympiad, which was held in South Korea in July 2000. The IMO is the pinnacle of excellence in mathematics for school pupils the world over. Every year, around 80 countries each send a team of 6 contestants to the IMO. There they sit two 4½ hour exams, each containing just three questions. Medals are awarded for good performances. This year the host country was South Korea, which has not hosted an IMO before.

Let us start with some of the events leading up to the IMO. The selection of the team started with the Senior Mathematics Challenge, a multiple-choice paper sat by more than 30000 students, taken in November 1999. The SMC lasts 90 minutes, and consists of 25 questions, of which the first 15 are meant to be widely accessible and the last 10 rather more testing. Based on their performance in the SMC, around 800 contestants proceed to the next round, the BMO1. This is a far, far harder paper, containing just 5 questions to be done in 3½ hours. Anyone who solves a BMO1 question has reason to feel pleased with himself/herself! The BMO1 is held in mid-January, and is followed by an amusing weekend in which 20 or so academics, teachers and ex-Olympians gather together to mark the scripts.

About 100 pupils qualified for BMO2, which is a still harder exam, consisting of 4 questions to be attempted in 3½ hours. Based on BMO2, 20 pupils were selected for the Trinity Training Session at Easter – these 20 included those whom we felt are realistic contenders for the team, and also some younger pupils who look like good prospects for the future.

The Trinity Training Session is an extremely intense and exciting experience for everyone. It lasts four days. For the first three days, the students have a variety of sessions, some taught to the whole group of 20 and some taught in groups of six or seven. The emphasis is on the students trying problems: the actual amount of 'lecturing' is kept to a bare minimum. The final day is, by contrast, rather different. The main event is the last of the selection exams, the Final Selection Test. The FST is designed to resemble a real IMO paper: there are just three questions, and the time allowed is 4½ hours.

In the next day or so, a squad of eight is selected. The choice is based on performance in FST, BMO2 and BMO1, and also on how the students have performed during the Training Session. The eight are notified within a few days of leaving Trinity, and they then embark upon the final and most gruelling part of the selection. This is the dreaded 'correspondence course'. Every week to ten days, the students are sent a sheet of about eight hard problems. They send in their solutions, which are marked by the Leader (myself) and the Deputy Leader (Richard Atkins, Head of Maths at Oundle). After about five rounds of this, the team of six is chosen, with the other two acting as reserves. Of course, the two reserves contribute immeasurably to the success of the team, as their presence during the training course has forced people to work hard for their places in the team!

In the week before the IMO itself, the team gathers at Birmingham, where the Summer School for younger pupils is held. As well as participating in some of the events of the Summer School, the team members receive some final training and preparation. This year, the squad of eight was as follows.

Team:

Thomas Barnet-Lamb	(Westminster School)
Stephen Brooks	(Abingdon School)
David Collier	(King Edward VI School, Southampton)
David Knipe	(Sullivan Upper School)
Michael Spencer	(Lawnswood High School)
Oliver Thomas	(Winchester College)

Reserves:

Hannah Burton	(City of London Girls School)
Kerwin Hui	(Berkhamsted Collegiate School)

Of these eight, Thomas and Ollie were 'returners' from last year's IMO in Romania (where each won a silver medal). Stephen and Michael were last year's reserves.

Next came the IMO itself. The IMO, for the students, was to start in Taejon on July 16th, but the Team Leaders flew in three days early, to select the questions that would be used. Each country has the right to submit some questions (months in advance); the host country then narrows these down to a short-list of about 25 questions, and it is from these that the six must be chosen. For those three days, the Leaders were kept in a secret location, far from Taejon – in fact, for obvious reasons, they were allowed no contact at all with the teams or the Deputy Leaders until the last exam had finished! This year, the Leaders met in Chonan, in a large complex just outside town that is normally used for training communications industry workers.

Some countries send an 'Observer' with the Leader or Deputy: this is usually someone who will do the job in a later year, and is coming along to see how things work. This year, however, things were very different for the UK, as we will be hosting the IMO ourselves in 2002. For this reason, we sent plenty of Observers. As Observers with Leader, there were Adam McBride, Duncan Harvey and Robert Smart, while as Observer with Deputy we had Alex Barnard.

The Jury chose the questions, and supervised the various translations. This year, 82 countries participated, which necessitated using more than 50 languages. The Jury consists of all the 82 Leaders: as one can imagine, a committee of this size functions in a rather chaotic way, but it does seem to reach sensible decisions.

Meanwhile, the team, led by Richard Atkins, had arrived in Taejon. After acclimatisation, and a pleasantly short Opening Ceremony, the actual exam dates were July 19th and 20th. There then followed a period of 48 hours of intense activity for the Leaders and Deputies. The Leaders move to a hotel in Taejon, and the Deputies join them there to help with the marking of the exams (a considerable increase in luxury from their previous accommodation with the students). Each country marks its own students' scripts, and then goes to 'coordination' for each question: this involves meeting with two Korean mathematicians and agreeing on marks. Finally, totals are worked out, and the cut-offs for medals established. The rough principle is that the ratio of Gold to Silver to Bronze to no medal should be very close to 1 to 2 to 3 to 6.

After a day of socialising, there was a Closing Ceremony, at which the medals were awarded. Everyone flew home the next day.

Now to the papers. Each paper has three problems, to be done in 4½ hours, with each question worth 7 points.

FIRST DAY

Problem 1.

Two circles S_1 and S_2 intersect at M and N. Let L be the common tangent to S_1 and S_2 closer to M than to N, and let L touch S_1 at A and S_2 at B. Let the line through M parallel to L meet the circle S_1 again at C and the circle S_2 again at D. Let the lines CA and DB meet at E; let the lines AN and CD meet at P; and let the lines BN and CD meet at Q. Prove that $EP = EQ$.

Problem 2.

Let a, b, c be positive real numbers such that $abc = 1$. Prove that

$$\left(a - 1 + \frac{1}{b}\right)\left(b - 1 + \frac{1}{c}\right)\left(c - 1 + \frac{1}{a}\right) \leqslant 1.$$

Problem 3.

Let $n \geqslant 2$ be a positive integer and let λ be a positive real number. Initially, there are n fleas on a horizontal line, not all at the same point. A *move* is defined as follows: choose any two fleas, at points A and B, with A to the left of B; let the flea at A jump to the point C to the right of B with $BC / AB = \lambda$. For each value of n, determine all values of λ such that, for any point M on the line and any initial positions of the n fleas, there is a finite sequence of moves that will take all the fleas to positions to the right of M.

SECOND DAY

Problem 4.

A magician has one hundred cards numbered 1 to 100. He puts them into three boxes, a red one, a white one and a blue one, so that each box contains at least one card. A member of the audience selects two of the three boxes, chooses one card from each and announces the sum of the numbers on the chosen cards. Given this sum, the magician identifies the box from which no card has been chosen. How many ways are there to put the cards into the boxes so that this trick always works?

Problem 5.

Determine whether or not there exists a positive integer n such that

n is divisible by exactly 2000 different prime numbers, and $2^n + 1$ is divisible by n.

Problem 6.

Let $A_1A_2A_3$ be an acute-angled triangle, with altitudes A_1H_1, A_2H_2, A_3H_3. The incircle of $A_1A_2A_3$ touches the sides A_2A_3, A_3A_1, A_1A_2 at T_1, T_2, T_3 respectively. Let the lines L_1, L_2, L_3 be the reflections of the lines H_2H_3, H_3H_1, H_1H_2 in the lines T_2T_3, T_3T_1, T_1T_2 respectively. Prove that the lines L_1, L_2, L_3 determine a triangle whose vertices lie on the incircle of $A_1A_2A_3$.

You are invited to send in solutions, enclosing an SAE please, to:

Imre Leader, Department of Pure Mathematics and Mathematical Statistics, Centre for Mathematical Sciences, Wilberforce Rd., Cambridge CB3 0WB.

The countries of origin of these questions were Russia, USA, Belarus, Hungary, Russia and Russia, respectively. Although none of the United Kingdom's problems were included, a remarkable three of them made the short-list: two by our veteran problem-setter David Monk (Edinburgh

University), and one jointly by myself and Peter Shiu (Loughborough University). For one person to have two questions on the short-list is extremely rare!

The team did very well, with everyone securing a medal. The total haul was two Silver and four Bronze. As a whole, the UK team's total score of 96 points (out of a possible 252) put us 22nd of the 82 countries. When looking at such results, it is important to bear in mind that our training programme is tiny compared with those of other countries (most of which have training camps lasting one or several months, and in some cases even a whole year!). Indeed, perhaps a more revealing statistic is that, among the Western European countries (where long training camps are the exception rather than the rule), we came 2nd, beaten only by Germany. Here are the top 10 teams, with their scores:

China 218, Russia 215, USA 184, South Korea 172, Bulgaria and Vietnam 169, Belarus 165, Taiwan 164, Hungary 156, Iran 155.

Here are our individual scores, with marks per question:

	Q1	Q2	Q3	Q4	Q5	Q6	Total	
Thomas Barnet-Lamb	6	0	0	7	7	1	21	Silver
Stephen Brooks	5	0	4	7	1	0	17	Bronze
David Collier	7	1	0	7	7	2	24	Silver
David Knipe	0	7	1	4	0	0	12	Bronze
Michael Spencer	2	2	0	6	1	0	11	Bronze
Oliver Thomas	7	1	0	1	0	2	11	Bronze

Four contestants scored full marks: two from Russia, one from China and one from Belarus. The cut-offs for medals were 30, 21 and 11 for Gold, Silver and Bronze respectively.

The IMO was a magical experience for the Leaders as well as for the Teams. For the Leaders, there was the wonderful initial three days of meeting as the Jury, all thinking and arguing about the problems and discussing different ways to solve the questions, then the two, unbelievably hard, days of marking and coordination, and then the last few days socialising with the team and other teams. For the contestants, there was the rather awe-inspiring arrival, being plunged among the best few hundred school pupils in mathematics from around the world, then the tension of the actual exams, then the sometimes equally great tension of waiting for the results, and in amongst all this the meeting with other teams from different parts of the world.

Perhaps the best way to convey some of the atmosphere of the IMO is to record my daily diary of some of the things that happened.

July 12: I meet up with Duncan and Robert at Stansted. Duncan is coming with a view to getting more involved with the role of Deputy – he would normally be Observer with Deputy, but that place is filled by Alex Barnard (who is coming along to see what the IMO is like, as possible preparation for IMO2002). Robert is one of the main people involved with the organisation of IMO2002, so he is coming to see how the IMO works in practice. The third Observer with Leader, Adam McBride, has left two days earlier, as he is on the IMO Advisory Board, a committee of about ten respected people who make sure that the IMO progresses successfully from year to year.

We change planes in Frankfurt, and it seems that our flight on to Seoul is the 'IMO flight', as I spot the Leaders from Belgium, Germany, Italy, Luxembourg and Spain on the same plane. We settle down for the long flight.

July 13: After some terrible in-flight movies, we arrive at Seoul around noon local time (Korea is eight hours ahead of the UK). Just before landing, I rush off to the toilet to put on gallons of mosquito repellent. Everyone laughs at me: although my doctor had told me to wear it 24 hours a day, it seems everyone else has been told that there is no problem as long as we are not out in the bush. (No-one has any problems with mosquitos, so it seems the majority was right.)

As we walk out of Customs we are astonished to see an array of TV cameras and bright lights. The IMO seems to be big news in South Korea. We head for the large IMO placard, and meet up with old friends from last year. Until that moment, we have not known where we will be going (for security reasons: the one thing we know is that it will *not* be Taejon!). We are told that we are going to Chonan, a medium-sized city about halfway to Taejon.

The air is incredibly hot and humid as we step across the tarmac to the coach. En route to Chonan, we stop to pick up the IMOAB, who have been meeting in Seoul but are now coming with us to Chonan, so we meet up with Adam.

The venue at Chonan is very impressive. It looks like something out of a James Bond film. A huge institute, in acres of grounds, with no other habitation anywhere near, has been given over to us exclusively. This is where the future Korean leaders in the field of communication are trained. There are about 1000 rooms, but it is being used just by the 150 or so of us (Leaders, Observers, and local organisers). No-one can believe the sheer scale of the place.

As we check in, along with our room keys we are given the shortlisted problems, minus solutions. We have about 24 hours to look at these problems and form our opinions of them before the solutions are given out. This is one of the best parts of the IMO. There are so many beautiful questions that one does not know where to start! It is utter joy as we go off to our rooms to have a think about some of them.

The time passes quickly, and soon it is time for dinner. This is a huge buffet, containing both Western dishes and, more interestingly, Korean ones. Most of the food seems very spicy (like Chinese food, but hotter), and rather delicious, particularly a beef stir-fry called bulgogi. This is something of a national dish, it seems, as we have it several times in our stay. Another national dish is kimchi, which is pickled cabbage with a very hot red sauce on it. This is served as an accompaniment to *every* meal, even breakfast.

After dinner, off to play table tennis with the Leader from Ecuador, who is well known for his Chinese-style serve. Then back to thinking about the problems. It is rapidly becoming clear that they are a *very* nice set.

July 14: Breakfast includes, as well as kimchi, most other possible kinds of vegetable (cauliflower, broccoli, carrots, tomatoes, cucumber). Plus plenty of eggs, bacon, sausages and so on, luckily. Then back to the problems. Adam is obsessed with a functional equation, and Duncan is looking at a three-dimensional shape-packing problem. I discover that the New Zealand Leader (new this year) and I get on well together, and we work together on several problems.

At 4.30 comes the first Jury meeting. After various unexciting speeches of welcome, we get on with our preliminary work. This consists in our throwing out questions that have been seen before (or that are very close to such questions). However, the Problems Committee (who had selected the short-list) have done a very thorough job, trawling through every possible book or magazine of problems, so only one problem on the short-list is rejected at this stage.

Dinner has lots of fantastic sushi, including some hard-to-identify things that turn out to be sea cucumber (delicious, but very chewy) and marinated squid (even chewier, so that one really has to 'chew to exhaustion').

The solutions have been given out before dinner, so everyone is keenly flicking through the booklet. Then more table tennis and also some drinking. There is a large computer room, and various Leaders who cannot live a day without email go off there.

July 15: Over breakfast, it turns out that several Leaders have been ill in the night. After some questioning, it transpires that they are exactly the Leaders who have drunk water direct from the tap. I am thankful for my bottled water.

Now comes a whole day of Jury meetings. We are starting off with a 'difficulty poll': each Leader is given a form on which to grade each problem on the short-list as easy, medium, hard or very hard. Then it is decided that we should have another column, marked 'good for IMO?'. We are all set to go when someone points out that we do not know what 'good for IMO' means. Does it mean 'suitable for an IMO question' or 'I like this question' or 'I would like this to be one of the six questions chosen'? There is much discussion, mainly by people saying it is 'obvious' what it means. We eventually go for 'suitable', and then it is suggested that we also mark six questions as being our favourites – note that this may well be different from the six questions we would like to see on the paper, for example because of balance.

Suddenly, a Leader requests that the Organising Committee should be making notes of what everyone is saying, as 'it is hard to remember afterwards'. I think 'Oh no, that is absurd, and will slow things down incredibly, but the Koreans are so polite that they will think they have to oblige'. Luckily, the Australian Leader immediately gets up and say words to the effect of 'What utter nonsense'. So the day is saved. In the coffee break, he is surrounded by people saying 'Well done'.

The results of the vote are anounced. There are five problems that stand out head and shoulders above the rest in popularity, and rather amazingly they are a beautiful spread of subjects and a good mixture of levels of difficulty. It seems an obvious choice to go for those five straightaway. But immediately several Leaders speak about 'caution … must not rush these important decisions … there are many other problems', so nothing is decided. (As it turns out, all five will later be chosen).

There is a long afternoon break, in which Adam goes ten-pin bowling with the New Zealand and Australian Leaders. The Australian Leader manages to fall. Not just a fall onto the wooden floor, but a fall onto the edge of the gutter that runs alongside the lane, so a nasty sharp edge. He is hobbling, and teased a lot, for the next few days.

That evening there is a banquet, hosted by the Regional Governor. Dish after dish is brought, and we lose count of the number of courses. Then a Korean orchestra plays, the highlight being some very loud drums, played violently yet carefully.

July 16: Over breakfast, several of us hatch a plan to get the five popular questions chosen as a package. It will save huge amounts of time, and it seems criminal to waste the opportunity presented by having popular questions that have good balance in subject and simultaneously in difficulty. So the plan is that the Finnish Leader will, right at the start of the Jury meeting, propose a motion that we adopt these five questions. Then the rest of us will chime in in support.

But, to our horror, the Chairman sees another hand raised first, and so calls on a different Leader, one who had not been privy to our breakfast-time plotting. He proposes that first we select two easy questions. (In the IMO it is traditional to have two easy, two medium and two hard questions, although of course this is very hard to judge, especially if one has not tried the question for oneself but merely read the solution. In any case all of these are relative terms – as none of the shortlisted questions are trivial!)

This seems to meet with assent around the room, so the moment to go for all five together has passed, and now we must get on with choosing in dribs and drabs. Two 'easy' problems are duly selected, one by one. The breakfast group who had wanted a quick resolution meet in the coffee break, and we are pleased that at least some progress has been made.

On we go. Someone makes a formal motion to adopt problem N3 (one of the five popular ones), but so many Leaders say 'wait, this may tie our hands in terms of the subject-matter of other questions' that he withdraws his motion. Then the Turkish Leader makes a plea about problem G8 (one of the least popular, with only one Leader putting it on his list of six favourites), saying that it would make a wonderful hard problem. He speaks very persuasively, and when he makes a motion and we vote the majority are in favour.

Well, except that it seems unclear what the voting rules are. For a motion to succeed, does it need a majority of the votes cast (for and against), or a majority of all votes (including abstentions)? There are different views, but it is decided that in the past the phrase 'absolute majority' has meant 'majority including abstentions' (although some Leaders have different recollections). So we are told that the motion did not pass after all. (In fact, this problem *will* end up being chosen.)

Someone now proposes a package of N3 and A5 (A5 is another of the favourites). Amazingly, given the earlier (utterly unfounded) worries about N3 by itself, no-one objects to this, and the pair are voted through easily.

The one remaining favourite that has not been voted through is A1, an inequality question. It is proposed, and there is now a discussion about whether it is really 'medium' or not. The trouble is that it has been placed at the top of the 'algebra and analysis' page (at position 1), denoting that the Problems Committee think it is the easiest of those questions. Now, it is certainly *not* easy, and in fact is pretty hard as inequalities go. But many Leaders just think 'Oh, it is number 1, so it must be easy', with the result that on the difficulty poll it has been overwhelmingly rated as 'easy'. So now there are some speeches about how A1 is actually not that easy. These seem to carry the day, and A1 is voted onto the paper.

Thus, after lots and lots of discussion, we have the five favourites after all. Now we just need one 'difficult' question and the paper will be finished. There are three or four leading candidates, of which the front-runner is probably the geometry question that we had almost voted in before. Another is a strange and unusual question about polynomials. At first I have no strong feelings about this question. But then the Hungarian Leader makes a speech pointing out that, as the question is so unusual, it is a very fair question: it does not give an advantage to over-trained teams. I am completely persuaded, and as he sits down I turn to the American Leader, who sits next to me (in between UNK and USA there should be URU, but the Uruguay Leader is sitting near the other Spanish-speakers), to say how much I agree. But he just says 'If you saw this problem, would you want to work on it?' And I realise that it is not a good IMO question after all! His point is that one should have appealing, attractive questions: questions that, if you heard them, you would want to tell your friends.

In general, the American Leader is quite amazing to sit next to. He came to America from Romania, and so is incredibly well versed in all things to do with mathematics competitions. He is full of wise comments, and I get the benefit of many whispered asides. We seem to agree on many things, and often support each other.

The geometry question runs out an easy winner. So now we have our paper. At this point the Problems Committee announce to us the countries of origin of the short-listed problems. We have not known them until now, to discourage 'political' voting. We are astonished at the Russian contribution: not only are *three* of the chosen problems Russian (a record), but all six of Russia's problems had made the short-list. And in fact each of the other three Russian problems had been strongly in the running to be chosen.

Next comes the preparation of the official English-language version. The ten or so Leaders from English-speaking countries retire to a quiet room to haggle over wording. The biggest problem turns out to be 'number of prime factors': does this mean with or without multiplicity? We go to incredible contortions to try to make it clear which one we mean. Then there are several arguments about things like 'which' versus 'that'. It is a typical gathering of mathematicians: pedantry abounds. Although at the same time everyone is doing their best to make things as clear to the contestants as possible.

Later that afternoon, we learn that our team has arrived in Korea. The Cubans have not been so lucky: their flight has been cancelled, so they must wait a few days for another one.

July 17: Now it is time for the English-speakers to relax, while the other official versions are produced (in French, Spanish, German and Russian). It turns out, however, that there is unhappiness about our English version. We (the English speakers) think 'How dare non-native speakers try to tell us how to write English?', but they do have a point. And the Georgian Leader comes up with a choice of wording that we had not thought of but is clearly superior to our own version.

After lunch, more relaxation, while all the translations are made. The computer room is filled with Leaders whose computer screens show a bewildering array of alphabets. There are about 50 languages needed!

July 18: The translations are pinned up for us to look at, and we wander around gazing at all of them. Some of them are very beautiful. Then, after lunch, off we go to the Opening Ceremony. This is in Taejon, so we go by coach. (It seems that coach journeys are an integral part of the IMO.) When we arrive at the auditorium, we have to wait without getting off the coach, because some contestants are still milling around outside. Only when they are safely seated are we allowed out. Some of those milling around are our team, so we manage to wave to them.

The ceremony itself is mercifully short. Impressively, the Korean Prime Minister is present, and gives the opening speech. Then some dancing, and a choir of young girls called the 'Little Angels' come on. They sing (very nicely) such traditional Korean songs as 'Banjo on my Knee' and 'Old MacDonald'.

Then back to Chonan. After dinner, a lot more table tennis, with the Leaders from Mongolia, Kuwait and Ecuador. The Ecuador Leader has found his special table tennis shoes, and this seems to make him very hard to beat.

Later that evening, we wander into the bar, to find some animated discussion in progress. It turns out that the final problem we had chosen bears some similarity to an IMO problem from 18 years previously (some Leaders have an encyclopaedic knowledge!). There is much argument about how similar it really is, and it is decided that we will discuss it in the Jury the next day.

July 19: This is the first day of the competition. For the first half-hour, the contestants have the right to ask questions. They must submit their questions in writing, and these are then faxed to us (as they are in Taejon while we are still in Chonan). When a question arrives in the Jury room, the relevant Leader reads out the question and proposes his response. The Jury discuss this and may alter the wording or even completely change the reply.

Most of the questions are pretty hopeless: questions whose answer is 'Read the question again'. Some cause great amusement, such as the student who asks 'What is a flea?'. His Leader proposes 'A small insect', but we decide that this does not give the right information, so we amend this to 'A small insect that jumps'. One important thing is to be consistent: if we answer one question with say 'No comment' then we must answer all identical questions with 'No comment' as well. This causes some heated moments, if a Leader thinks his student has been unfairly treated.

After the question-and-answer session, we turn our attention to the similarity issue about the final question. But the overwhelming view is that the similarity is very slight, so it is decided that the question will not be changed.

Now is the time for the mark-schemes to be introduced and discussed. The Chief Coordinator comes to tell us the proposed schemes. (Each question has six Coordinators, who work in pairs, plus a Senior Coordinator who moves between the pairs during coordination. And overseeing all of them is the Chief Coordinator.) Some of the mark-schemes look rather vague, and there is much heated discussion. In particular, it seems as though the magician question has a strange mark-scheme: in their efforts to give a few points to people who have just done a bit, the setters have ended up penalising people who have got the question right but with only minor mistakes. In particular, it seems that one can get up to 5 points without having the key ideas, whereas one slip in an otherwise fully correct proof can lead to a score of only 3.

In the afternoon, we are taken to a folk museum: a recreated old village (this may be an IMO tradition, as in the 1999 IMO we did the same thing on the same day). Lots of dancers, plus a spectacular tightrope walker, who at one point shuffles along on his knees. A camera crew ask me to say something 'short, very short' about Korea. I say something like 'Korea is wonderful, people very friendly, sushi delicious', and they say 'Yes, thank you, but shorter, please'.

Back at Chonan, we are waiting for the first day's scripts to arrive. They are due around 10pm, the lateness being because every script must be photocopied before we see them (so that the Coordinators can have a look in advance). But there is a hitch. It seems that two students have inadvertently taken their scripts out of the exam room in their backpacks. The organisers do not want to start photocopying until all the scripts are in one place, which involves sorting this matter out. It turns out that there is no suspicion of foul play, so the two scripts are added to the general pile and photocopying can begin. The new estimated arrival time is midnight.

Our rooms have an intercom system connected to the main reception desk, so we are promised that as soon as the scripts arrive we will be summoned from our rooms. Sure enough, at a quarter to midnight we learn that the scripts have arrived. Everyone rushes to pick up their scripts – the reception area is a sea of people and chatter.

I find a corner and sit down to have a look. I start with the geometry problem, as it is the one I expect the team to have done the worst on. But we have solutions from Thomas, David C and Ollie. Plus a nearly-solution from Stephen. He has clearly been under time pressure, as his work ends with a proof that three triangles are similar and a statement that 'the result ought to follow from this'. It does, easily, and with one more minute he would have done it. But he will only lose a point or two for this. Four solutions: I am absolutely delighted.

Then I look at the inequality, which we ought to be good at. And my mood changes. David K has solved it, with a highly original and beautiful proof. But no-one else has. I cannot believe it. Michael has made some progress, but not much.

Finally, I look at the fleas question, and this is even worse. There is only one solution offered, by Stephen. And, on closer inspection, this proof has a fatal flaw. The team have also disobeyed a key instruction: to always hand in all one's rough work. Either that, or they have spent almost no time thinking about this question, as almost no work has been handed in. The actual answer is easy to guess (although proving it correct is the whole meat of the question), but apart from Stephen only David K has even done that. I am devastated.

Robert has stayed up as well, and we try to find out how other teams have fared. We are next to the American Leader, and soon find out that his team has done well – he is quite cheerful.

July 20: The second day of the contest. Another question-and-answer session. Some amazing questions: one student asks, about the magician question, 'Is the magician told which boxes the two cards have come from?'! We decide that the only possible answer is 'Read the question again'. Actually, this is almost certainly more helpful than just saying 'No', because to even ask such a question the student must have made a severe error in his reading, so we do need to signal this to him. One student asks about the number theory question so many times that it is suggested that we should send the reply 'Read the question for the fourth time'.

Then we are transferred to Taejon. We settle into our hotel, where the Deputies will be joining us to help with the marking. Then we take a taxi to the competition site, to greet our team when they come out from the exam. We see Richard and Alex, and they tell us that the team had real problems on the first day from lack of sleep: the rooms are not air conditioned, and the humidity is very high.

Out come the team, relieved to have finished. We go to lunch with them, and discover that their food is surprisingly good. In addition to the nice main courses, there are unlimited soft drinks and unlimited ice cream.

Richard and Alex come back with us to the hotel, and we look at yesterday's solutions a bit more. Then the day's scripts arrive, and we rush off to get them. We tear open the folders. Thomas, Stephen, David C and Michael have all done the magician question. Each has done it in his own inimitable style. Thomas has found a wonderful induction way of doing the question, David C has the 'standard' solution, but in very short form, Stephen has a variant on this and Michael has an extremely clever, very unusual proof. Our spirits rise.

Then on to the number theory question. A long and ingenious solution from Thomas (so strange that, to this day, I have no idea how he thought of it!), and a short elegant solution from David C. Michael has some clever ideas, but there is an error in his proof, and almost nothing can be salvaged.

The geometry question, Problem 6, is the disaster we had feared: no-one has really got anywhere on it. So, all in all, we are not very happy. But now we must hurry up and prepare for the coordination sessions. These will start tomorrow morning. We go and look to see what our times will be. For each Problem, there are three columns on the noticeboard, giving the times for each team (each column corresponding to one pair of Coordinators). It takes us ages to work out the system the organisers have used. They have tried to put all the top teams together with the *same* pair of Coordinators for each question – clearly sensible for consistency. But this would overload the timing: there is half an hour per team per question, and of course the good teams tend to have more answers, more written, more to discuss. So in among the good teams are a few of the not-so-good teams, just to speed things along! This is all extremely sensible. We are in the top group, and it seems that for alphabetical reasons we will be followed into almost every coordination by the Americans.

July 21: We are starting with a bang: 9am is coordination time for us for Problem 6. Actually, we are quite lucky in this, as we don't have much to argue about, so this will probably be the easiest coordination session of all. Of course, however, every point is vital.

Richard, Duncan and I gather outside the room where Problem 6 is being coordinated. The rules are that only two of us may speak during coordination, so Duncan will be the non-speaking person (we always have to ask permission to bring in an Observer, but this is always granted). In fact, I will also be a more-or-less non-speaker, as it is Richard who has gone over the scripts in detail for this problem. I am very nervous – it is that butterflies-in-the-stomach feeling that everyone who has waited to be called in to an important meeting must know.

We are called in, shake hands with the two Coordinators, and off we go. One of the Coordinators is Marcin Kuczma, who had been the Polish Leader the previous year (the Koreans imported two foreigners to help with problem selection, and also act as Coordinators). He has become a good friend of mine, so it is strange to be facing him across the coordinating table! Ollie and David C have made a bit of progress, and get 2 points each. Thomas has done less, and gets 1 point, which is fair. The rest get 0. So this has gone as expected.

Back to our room we go, to prepare for Problem 2. There may be some trouble with David K's solution. He has two 'without loss of generality's. The first is fine, but the second should come earlier than he has placed it. This is not a problem, as later he has inserted something saying 'oh, please move that WLOG to the relevant place'. But the Coordinators may not have seen that, so we may have to show it to them.

The other script that we are spending time on is Michael's. He has made some preliminary reductions, and has given some vague 'growth speed' reasons as to why what he is left with to prove is true. This will score a maximum of 2 marks. But it occurs to us that 'growth speed' suggests calculus, so that if his proof can be finished off by calculus we can at least say 'look, here is how to make it into a proof'.

I try to do it by calculus. One gets a horrible expression that one has to show is always negative, but I can see no reason why it should be. So I ask Alex to have a look, and in a few minutes he has come up with a clever argument that establishes it. It is for things like this that it is so nice to have plenty of people helping with the marking. Alex, Adam and Duncan all give lots of help to Richard and myself – it is quite a luxury to be able to say to someone else 'look at this script, please'!

Straight after lunch we are on for Problem 2 coordination. For David K's script, the Coordinators say that they have a problem with the 'without loss of generality'. We try to point out where the error is corrected, but they seem not to understand. It eventually turns out that we have been talking at cross purposes: they are happy about the second WLOG, but it is the first one they want explained. We explain it, and David gets his 7.

When we get to Michael's question, we ask whether a continuation by calculus would be grounds for an extra point for Michael's 'growth speed' nonsense. No, we are told. And this is no surprise, to be honest. So Alex's nice argument never sees the light of day.

Now we have a couple of hours before our final coordination of the day, which will be on the dreaded Problem 4 – the one with the silly mark-scheme. Thomas, Stephen and David C will get 7. There is a gap in Michael's proof, but it is easy to fill (not needing any new ideas), so he will score 5 or 6. The one that will be difficult is David K's script. His neat work is nonsense, but in his rough work there are plenty of ideas. In fact, his rough work contains all the ideas one needs for the proof, but in a strange jumbled-up order. Plus there is one case that he has failed to consider. In a sensible mark-scheme, this would be worth 4 points, but this mark-scheme will only give him 2 or 3, depending on how it is interpreted.

As we wait to go in, I am finding it very hard to keep calm, because I hate the mark-scheme so much: it is so unfair. Anyway, we start off easily enough: 7 for Thomas, 7 for Stephen, 7 for David C. Then we get to David K's solution. The Coordinators offer 2 points, and we explain why we want 3. They stick to 2: they are not being at all flexible. I try to explain that the mark-scheme could be interpreted as giving 2 or 3 (depending on what one counts as 'significant progress'), but that, since the whole scheme is biased towards people who have enumerated a few cases and against people who have tried to solve the problem, we must interpret the scheme in favour of the latter class of people. The Coordinators listen and say 'Yes, we understand what you say, but we are offering 2 marks'.

After some more heated discussion, I sense that we are getting nowhere. It is terribly unfair that David's attempt is not getting the credit it deserves, but I suspect that if we go on arguing then the Coordinators will just say 'well, this is how we are treating everyone', and of course that cannot be argued with. So, to speed things up, I say 'OK, I am prepared to sign if you tell me that you have had similar scripts before that you also gave 2 points to'. (Signing means signing the mark-list, which is the official endorsement that I and the Coordinators agree to the given marks.)

The Coordinators produce a script and invite me to read it. It is a Russian script, a beautiful solution, with one case overlooked (and remember that this mark-scheme savagely punishes overlooked cases). I say that I imagine this is worth 6 points, and they say 'Well, we gave this 3 points'. I say that that is terrible, but that of course that solution deserves far more points that David's, so I am sadly ready to take the 2 points.

Then Svetoslav Savchev (who is from Bulgaria, and is the other foreign Coordinator) says 'The Russian Leader refused to sign, by the way'! So I also refuse to sign, and we carry on arguing. By now our half-hour is nearly up, and we are getting nowhere. Again I sense that eventually the 'this is how we have decided to do it' argument will carry the day, so I say 'OK, I will sign, but with heavy heart'. I am just reaching out to put the script away when Svetoslav says 'Remember, once you sign, that is the point of no return'. He is essentially telling me to refuse to sign! This is great evidence that the Coordinators' role is not just adversarial. We arrange to come back at 9pm that evening.

Over dinner, I chat to the Russian Leader, who is outraged about his student's script, and also to the American Leader, who has a similar story to tell. Dinner calms us down a bit, as it is so nice: lots of sushi, including some incredibly soft and delicate tuna, and also a large dish of snails (cooked, not part of the sushi).

When we come back at 9pm, we discover that there is a queue of Leaders who are all coming back after refusing to sign for Problem 4. The 8.30 return slot is overrunning, so we are asked to come back the following afternoon instead.

We go back to our room to prepare for the next day. Problem 5 is ready for coordination, the only question being whether Michael will score 1 or 2. It certainly deserves at least 2, but I am afraid the mark scheme will be interpreted to give only 1. And Problem 3 is easy to prepare as well, as the only script that needs attention is Stephen's. Meanwhile, the foursome of Alex, Richard, Adam and Duncan are working hard on the Problem 1 scripts. They have discovered a slip in Thomas's solution. It has been caused just by a typographical problem (maybe misreading his own writing), but it is definitely a gap. Of course, if one had pointed this out to him then he would have instantly fixed it, but the question is what what he has actually written is worth. After a huge amount of agonising between 4, 5 and 6 points, the foursome decide that 5 is the right mark.

At midnight, I decide to have one more look at David K's Problem 4, just to make sure I am still on top of it when we go back to argue the next day. I have been perplexed by the fact that there is so much good stuff in his rough work, while his neat solution is such rubbish. Perhaps he thinks, after his rough work, that he *has* solved the problem, and then he thinks (mistakenly) that he has found some amazing 'short cut', to make the neat solution so short. So I decide to go back over his rough work and see if I cannot actually find a proof in there. And, after lots of reading, something amazing emerges. The rough work is full of paragraphs that seem to end in strange ways. If one *completely* ignores all those paragraphs, and then follows various signposts David has written like 'Now go to foot of previous page', then one does obtain, well, not quite a proof, but a proof with a gap. Elation! We decide to ask for 4 points on the basis of this.

July 22: The day starts with coordination for Problem 3. The Coordinators offer 2 points for Stephen's failed proof. But we explain that actually what his attempt does prove is rather more than the work for which the mark-scheme gives 2 points (although it is far from a solution to the problem), and they quickly agree to give 4 points.

After lunch it is time for Problem 5. As expected, Michael's script gets only 1 point, which I still feel is unjust (although it is consistent with the mark-scheme). The Coordinators are as impressed as we are with David C's super-short solution and Thomas's how-did-he-think-of-it solution.

Now comes the moment we have been waiting for. We are back for Problem 4. The Coordinators start by saying that they are sorry, but they are still only prepared to give 2 points. I say that I apologise enormously for what I am about to say, but I must up my request from 3 to 4. They are intrigued, and enjoy it when I explain the contorted way one has to read the script to see the proof-with-gap. After conferring, they say 'Yes, 4 points'. We are delighted.

Then we deal with the rest of the Problem 4 scripts. We come to Michael's script, and they say 'There is a gap. What mark do you propose?'. This shows again how carefully they have done their homework, reading hundreds of scripts in advance. We say 6, and they say that that is what they had thought as well.

Now there is only Problem 1 to go. We take a break to go and look at the huge scoreboards where the results of each contestant are being displayed. Several people are making charts and tables to try to predict the medal cut-offs. Of course, this is very inexact, because of the partial nature of the information. In addition, many teams have taken postponements on coordinations. This is often for the better teams (as there is more for them to discuss), and trying to compensate for this in the estimates is very difficult. Anyway, the Irish Deputy and the Dutch Deputy are keeping count, and they seem to think it will be 13 for Bronze, 22 or 23 for Silver, and 30 to 32 for Gold. This is indeed the impression most Leaders have.

We go back to our room to take stock. David C has 17 points so far, with 7 to come from Problem 1, so he is a safe Silver. Thomas has 15, with 4 or 5 to come, so a high Bronze. Stephen has 12, with about 5 to come, so this is a safe Bronze. Michael and Ollie seem to be out of medal contention: Michael has 9 so far, with 2 to come, while Ollie has 4 with 7 to come. So these two will narrowly miss out. And David K has 12 so far, so just 1 point from Problem 1 will give him a Bronze. Unfortunately, his script looks depressingly like a zero. He has stated that a certain quadrilateral is cyclic (which is relevant), but has not written down a proof of this! This looks like it will cost him a point. We are devastated. All that hard work over his points on Problem 4 in vain!

So in we go for Problem 1 with heavy hearts. The Coordinators ask us how many points we want for Thomas's solution, and we say 5, but they say 'We thought 6, actually'. So that is nice. After 5 for Stephen and 7 for David C, it is time for David K. Richard argues persuasively for 1 point, but the Coordinators say that they cannot give a point without an argument on the page for why this quadrilateral is cyclic. Richard argues some more. We are almost begging for this point, as it seems it will make the difference between medal and no medal. But we are bound to lose the argument, and indeed we do. We go out of the coordination room rather sad. When we see the Coordinators at tea, we apologise to them for wasting their time arguing about that lost cause, and we explain why it had been important to us. They say 'Oh yes, we guessed that'!

After dinner, the excitement is building, as everyone is clustered round the scoreboards. The predictions are now 13 for Bronze, 22 for Silver and 31 for Gold. This means that the team will get one Silver and two Bronzes, with the other three narrowly missing a Bronze and Thomas just missing a Silver. We feel terrible for the team – four near-misses.

But, as the late results trickle in, it seems that the postponed sessions have not led to such high marks as the experts had thought. By 9pm, it is certain that 21 will be enough for a Silver, so Thomas has done it. And by 9.30 the cut-off for Bronze has come down to 12. This is great – our work on David K's Problem 4 was not in vain after all. We are still sad for the other two, who will miss out on a medal by one point.

Then, with only about 20 contestants' scores incomplete, a rumour starts going round that maybe 11 points will be enough for Bronze. For this to happen, we would need half of all contestants to score 10 or less.

Suddenly we are galvanised into action. The first question is how many contestants there are in total: some people say 461 and some say 463. Alex and I decide to count it ourselves. We rush up and down the scoreboards, getting in everyone's way, and come up with 461. So we need to find 231 people scoring 10 or less.

We count 226 confirmed scores of 10 or less – meaning either completed scores of 10 or less or scores of 3 or less with one mark to come. The big grey area is the people with scores of from 4 to 10, with a mark to come. One of these is updated, and to our joy (although not the joy of that country's Leader, of course) the score is under 10. So up to 227.

We see the Czech Leader, and ask him about him one blank: a student of his on 5 with one mark to come. He says that yes, they are still fighting over his mark for one problem (Problem 4, of course), but that the fight is over whether it should receive 3 or 4 marks. So now we are on 228.

Then comes a period of a few minutes in which some scores are updated, but none in our favour: people being pushed over 10 points by their last mark. Eventually, by 10.30pm, there are just five contestants left who could affect things, and we are still on 228. Three of these are from Macedonia. So we frantically look for the Macedonian Leader, and ask him about those postponements. We know he has power of life and death over us. He asks why we want to know about his students, and we breathlessly explain why. And he tells us that none of the three has any chance of getting to 11 points. I kiss his hand (an event unfortunately captured on film by the Dutch Deputy). And off we go to the bar to celebrate. Everyone has got a medal (for the first time since 1996).

July 23: The day starts with our final Jury meeting. First there are two contestants whose marks on one question have still not been agreed (this is extremely rare), so we vote on these. Then we ratify the cut-offs. The Gold cut-off is officially 31, as to allow 30 would mean 39 Golds, which is very slightly above one-twelfth of 461. But it is so close (461/12 being about 38.5) that we decide to allow the extra half-person, bringing the cut-off down to 30. At last everyone can relax.

In the afternoon I go shopping with the New Zealand Leader and Deputy and the Canadian Deputy. And this is my introduction to bargaining. In one shop I buy a fan for 3000 won (about 2 pounds). When I come out, my companions ask me if I bargained. I say no, and they laugh and explain that one should always bargain.

So in the next shop, when I need to buy a disposable umbrella (against the sudden torrential downpour), I communicate via counting-by-showing-fingers to reduce an umbrella in price from 900 won to 700 won. I try to pay, and the assistant shows that I have not paid enough. Of course, I am out by a zero, as I have actually gone from 9000 to 7000.

I watch my companions bargaining, and feel I have absorbed enough to be an expert. So, in the next shop, when a price of 15000 won is given (for an *extremely* nice fan), I start to bargain. But the assistant is not willing to give an inch, and says 'No bargain'. I decide that this must be a hard case, and persevere. Then she says, in perfect English, 'We are not a bargaining shop'. It seems that certain shops are designated as non-bargaining. I feel very embarrassed, and am laughed at even more by my colleagues.

On the way out, we meet the Russian Leader, looking for shoes. His Deputy has brought his black suit from Russia, instead of his white one, and he only has his white shoes with him. He has spent half the day looking for black shoes, but without success, as no shoe will fit him – Korean shoe sizes are rather small.

After dinner there is a 'Korean Cultural Evening', to which the teams have been invited as well. We meet up with ours, and congratulate them. Then comes a fast and furious performance, with the highlight being some more loud and skilful drumming.

July 24: In the morning the team come to visit us at our hotel. We go onto the roof (illegally, I suspect) to admire the view. Then off they go to change into smart clothes for the Closing Ceremony.

The Ceremony is great fun. The South American teams give huge cheers when one of then goes up on stage to receive a medal. There is already a party atmosphere. Then back to the students' accommodation, which is where the Closing Dinner will be. We want to take photographs of the team in their smart clothes and also in their team T-shirts. We tell the team to pick up their T-shirts from their rooms but to keep their smart clothes on for the moment. It takes 10 minutes for them to understand this, and many of them change at the wrong time or into the wrong clothes, but eventually it is all sorted out.

Then the outdoor banquet, with a beautiful ice sculpture the focal point. After dinner some very loud music starts. The DJ invites contestants on to the stage if they would like to sing. After some harmless attempts, one contestant starts to sing a song with highly unpleasant lyrics. The DJ cuts off his microphone and says to the audience 'That is censored, so please forget what you just heard'. I set off to find YUG3 and ROM6. These are two contestants who are famous among the Leaders, as they are the only ones who have managed to solve Problem 6 but not Problem 1 (which seems impossible, as they are on similar topics but Problem 6 is so much harder). Eventually, with the help of their Leaders, I find them, and shake their hands. I also find USA6, who has managed to do Problem 6 by coordinates!

We head back to our hotel, and the bar. The team are staying up late, as the Americans are leaving at 4am and our team want to stay up with them.

July 25: A bus to the airport, and a long flight to Frankfurt. We say goodbye to Adam, who is flying to Edinburgh, and we fly off to Heathrow. We arrive at 10pm, exhausted but happy.

IMO2000 was extremely interesting, both for the mathematics and for the opportunity to see what Korea is like. The whole event ran very smoothly, which is a real credit to all the organisers. This is particularly impressive in view of the fact that Korea has never hosted an IMO before. We are very grateful to all of the Koreans for this fantastic experience.

Closer to home, I would like to thank

- all the pupils who took part in any stage of the UK competition
- all the teachers who encouraged them
- Peter Neumann, chairman of the UKMT, and Adam McBride, chairman of the BMOC
- Alan West and Brian Wilson, organisers of BMO1 and BMO2 respectively
- everyone involved with the BMO1 marking, particularly Christine Farmer and Brian Wilson
- Christine Farmer for help in the hectic period between BMO2 and the Training Session
- all those who helped with the Trinity Training Session, particularly Julia Gog
- all of our sponsors, particularly Trinity College, Cambridge for hosting the Training Session and the Royal Society for hosting our September celebration
- the DfEE for a grant covering travel to and from the IMO
- Ben Meisner for producing the BMO 2000 booklet, copies of which are now owned by practically every Leader and Deputy
- Richard Atkins for assistance with the Correspondence Course
- Richard Atkins, Alex Barnard, Duncan Harvey, Adam McBride and Robert Smart for all their help in Korea.

Perhaps most of all, I would like to thank Tony Gardiner and Adam McBride for continuing to give me advice on the role of Leader. Whatever I ask them, I always receive thoughtful and useful replies.

This just leaves the squad of eight. I have found them an amazing squad to work with. All are extremely bright and enthusiastic. Each of them has come up with novel and exciting answers in the Correspondence Course. I have particularly enjoyed the pleasure when one of them starts off a proof in a way that I 'know' will not work, yet manages to make it work! It has been wonderful to deal with them.

Hannah Burton and Michael Spencer will still be around next year. Of the others, two are off to Oxford: Stephen Brooks to read Maths at Trinity College and Ollie Thomas to read Classics at New College. The other four, Thomas Barnet-Lamb, David Collier, Kerwin Hui and David Knipe, will read Maths at Trinity College, Cambridge. We wish them all the very best for the future.

Other aspects of the UKMT and other similar bodies overseas

As well as the competitions described in the foregoing pages, the UKMT has produced other items.

Newsletter

Since September 1997, the UKMT has produced a twice yearly Newsletter called Maths Challenges News. This includes "behind the scenes" photos and articles. Among other things, it helps readers to put names to the faces of the people who make up the UKMT, and find out what they do. Photos of pupils who have taken part in Challenges, sent in by schools, are also published, as are occasional in-depth discussions of particularly interesting competition questions, and news about the UKMT's future plans. This year also saw two special crosswords included.

Web Site – www.mathcomp.leeds.ac.uk

In September 1998 the UKMT web site was launched on the World Wide Web. The site gives information on the organisation and its activities and also provides specially illustrated questions from previous competitions and links to other interesting sites which provide resources for young mathematicians.

Poster

In December 1998 the UKMT published its first poster, a colourful design titled 'Fermat's Last Theorem' showing in a fun, eye-catching way the development of the mathematical theories underlying this famous theorem from Pythagoras through to the proof by Andrew Wiles in 1994. The idea for the poster came from Dr A.D. Gardiner, and it was drawn by mathematician and artist Dr Joanna Brown. The poster was supplied free of charge to all schools on the Maths Challenges mailing list to enliven their classroom walls.

Leaflets

In July 2000 the UKMT published a new promotional leaflet for distribution to conferences, exhibitions and in response to enquiries from individual schools. It is hoped that the leaflet will whet the appetite of schools who do not currently participate in the Challenges, and give sufficient information for them to register their interest.

Overseas bodies

The UKMT has links of varying degrees of formality with several similar organisations in other countries. It is also a member of the World Federation of National Mathematics Competitions (WFNMC). What follows is a brief description of some of these other organisations. Some of the information is taken from the organisations' web sites but a UK slant has been applied.

Canadian Mathematics Competition

www.math.uwaterloo.ca/CMC/
CMCHome.html

The Canadian Mathematics Competition has established itself from the first contest in 1962, with around 350 Ontario entrants, to the current level of over 200,000 participants from more than 3,500 high schools across Canada. Over the years the competition, which is based at the University of Waterloo in Ontario, has grown to become a series of mathematics contests spanning a wide age range. Participants in several other countries also take versions of the papers.

Just as in the UK, the first challenge is a multiple-choice paper and there are follow-up papers for high scorers. In Canada these are named after famous mathematicians, such as Pascal, Cayley and Fermat. In Great Britain, the follow-up papers to the Cayley and Fermat papers appear as the International Intermediate Invitational Maths Challenges for high scorers in school years 10 and 11 (equivalent to the Canadian grades 10 and 11).

The contest operations involve hundreds of volunteers, including teachers from every province, consultants from various universities, and observers from different countries, who aid in such areas as contest creation and contest marking. The Chairman of the UKMT Problems Group, Howard Groves, took part in the creation of the Pascal contest in 1999.

“Kangourou des Mathématiques”

European Kangaroo Competition

www.mathkang.org/

The obvious question is: why Kangaroo? The name was given in tribute to the pioneering efforts of the Australian Mathematics Trust. The Kangaroo competitions, launched in 1991 by the "Kangourou Sans Frontières" organisation in Paris, have enjoyed similar huge growth and success. In 1999 a staggering 1.5 million candidates participated: over 500,000 in France and the rest throughout Western and Eastern Europe.

2,000 children in the UK in the equivalents of English Year 9 took part in the "Cadet" level of the Kangaroo competition as a follow-up to the Intermediate Maths Challenge. Two members of the UK Mathematics Trust, Mary Teresa Fyfe and Andrew Jobbings, took part in the setting of papers for the 1999 competitions.

The “Kangourou Sans Frontières” organisation produces a range of lively mathematical journals, books and posters with the highly characteristic and eye-catching ‘tessellating kangaroos’ created by the artist Raoul Baba.

As with all the competitions described in this section, the main objective is to stimulate and motivate large numbers of “average” pupils and contribute to the development of a mathematical culture which will be accessible to, and enjoyed by, large numbers of children and young people.

The Australian Mathematics Trust

www.amt.canberra.edu.au/aboutamc.html

For the past twenty two years the Australian Mathematics Competition has established itself as one of the major events on the Australian Education Calendar, with about one in three Australian secondary students entering each year to test their skills. That's over half a million participants a year.

The Competition commenced in 1978 under the leadership of the late Professor Peter O'Halloran, of the University of Canberra, after a successful pilot scheme had run in Canberra for two years.

The questions are multiple-choice and students have 75 minutes in which to answer 30 questions. There are follow-up rounds for high scorers.

In common with the other organisations described here, the AMC also extends its mathematical enrichment activities by publishing high quality material which can be used in the classroom.

Whilst the AMC provides students all over Australia an opportunity to solve the same problems on the same day, it is also an international event, with most of the countries of the Pacific and South-East Asia participating, as well as a few schools from further afield. New Zealand enters over 30,000 students, while Singapore enters a further 15,000 students to help give the Competition an international flavour. In fact 35 countries participated in 1998.

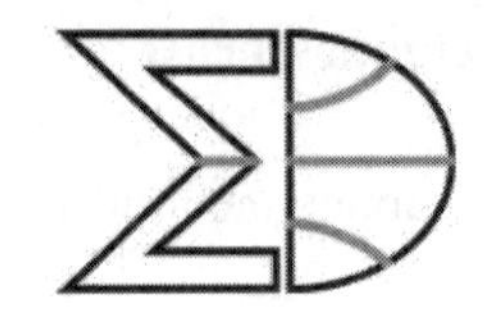

World Federation of National Mathematics Competitions – WFNMC

www.amt.canberra.edu.au/amtwfnmc.html

The Federation was created in 1984 during the Fifth International Congress for Mathematical Education.

The Federation aims to provide a focal point for those interested in, and concerned with, conducting national mathematics competitions for the purpose of stimulating the learning of mathematics. Its objectives include:

- Serving as a resource for the exchange of information and ideas on mathematics competitions through publications and conferences.
- Assisting with the development and improvement of mathematics competitions.
- Increasing public awareness of the role of mathematics competitions in the education of all students and ensuring that the importance of that role is properly recognised in academic circles.
- Creating and enhancing professional links between mathematicians involved in competitions around the world.

The World Federation of National Mathematics Competitions is an organisation of national mathematics competitions affiliated as a Special Interest Group of the International Commission for Mathematical Instruction (ICMI).

It administers a number of activities, including

- The Journal *Mathematics Competitions*
- An international conference every four years. Previous conferences were held in Waterloo, Canada (1990), Pravets, Bulgaria (1994), Zhong Shan, China (1998). The next, is scheduled for Melbourne, Australia in 2002.
- David Hilbert and Paul Erdős Awards for mathematicians prominent on an international or national scale in mathematical enrichment activities.

The UKMT sent two delegates, Tony Gardiner and Bill Richardson, to the WFNMC conference in Zhong Shan in 1998 and provided support for several delegates who attended ICME 9 in Tokyo in August 2000, at which the WFNMC provided a strand.

List of volunteers involved in the UKMT's activities

The Council

Dr Peter Neumann (Chair)
Dr Roger Bray (Secretary)
Professor John Brindley
Mrs Mary Teresa Fyfe
Mr Howard Groves
Mr Terry Heard
Miss Susie Jameson
Dr Imre Leader
Professor Adam McBride
Mr Dennis Orton (Treasurer)
Mr Bill Richardson
Professor Chris Robson
Dr Alan Slomson
Miss Patricia Smart
Mr Robert Smart

Trustees who are not on the Council

Dr Tony Gardiner
Mr Peter Thomas
Professor James Wiegold

The Subtrusts

British Mathematical Olympiad Committee

Professor Adam McBride (Chair)
Dr Tony Gardiner
Dr Imre Leader
Dr Alan Pears (Treasurer)
Miss Patricia Smart (Secretary)
Dr Brian Wilson

The Senior Challenges Subtrust

Mr Bill Richardson (Chair)
Mr Dennis Archer
Mr Colin Dixon
Miss Patricia Smart (Treasurer)

Junior Olympiad Subtrust

Professor Chris Robson (Chair)
Mr Dennis Orton (Treasurer)
Mrs Mary Teresa Fyfe (Secretary)
Dr A K Jobbings
Mr Bill Richardson

Junior and Intermediate Challenges Subtrust

Professor John Brindley (Chair)
Mr Howard Groves
Dr Andrew Jobbings
Professor Chris Robson
Dr Alan Slomson (Secretary and Treasurer)

Other Groups

Additional members of the BMOC Extended Committee

Richard Atkins (Oundle School)
Christopher Bradley (Clifton College)
Jason Brooks (Clifton College)
Kevin Buzzard (Imperial Coll. London)
Philip Coggins (Bedford School)
Tim Cross (KES, Birmingham)
Vin de Silva (Trinity College, Camb.)
Christine Farmer (Royal Holloway Coll.)
Julia Gog (Trinity College, Camb.)
Ben Green (Trinity College, Camb.)
Gerry Leversha (St Paul's School)
David Monk (ex Edinburgh University)
Peter Shiu (Loughborough University)
Alan West (ex Leeds University)

BMOC Markers

Dennis Archer (Bedales School)
Richard Atkins (Oundle School)
Richard Bridges (KES, Birmingham)
Philip Coggins (Bedford School)
Ed Crane (Trinity College, Camb.)
Michael Davies (Westminster School)
Vin de Silva (Trinity College, Camb.)
Christine Farmer (Royal Holloway Coll.)
Tony Gardiner (Univ. of Birmingham)
Ben Green (Trinity College, Camb.)
Howard Groves (RGS, Worcester)
John Haslegrave (Trinity Coll. Camb.)
Imre Leader (Univ. College, London)
Gerry Leversha (St Paul's School)
Adam McBride (Univ. of Strathclyde)
Ben Meisner (Oundle School)
Peter Neumann (Oxford University)
Sylvia Neumann (Open University)
Patricia Smart (Dean Close School)
Amanda Turner (Trinity Coll. Camb.)
Brian Wilson (Royal Holloway College)
William Wynne-Willson (ex University of Birmingham)

Markers for IIIMC and JMO

Roger Bray	(Royal Institution)	IIIMC
Dean Bunnell	(Queen Elizabeth GS, Wakefield)	IIIMC / JMO
Margaret Burrow		IIIMC
Stephen Campbell	(Merchiston Castle School)	IIIMC
Katy Chisholm	(Hutchesons' Grammar School, Glasgow)	IIIMC
Barbara Cullingworth	(St Bernard's Convent School , Slough)	IIIMC
Mary Teresa Fyfe	(Hutchesons' Grammar School, Glasgow)	IIIMC
Howard Groves	(Royal Grammar School, Worcester)	IIIMC / JMO
Terry Heard	(ex City of London School)	JMO
Andrew Jobbings	(Bradford Grammar School)	IIIMC / JMO
Gerry Leversha	(St Paul's School)	JMO
Adam McBride	(University of Strathclyde)	IIIMC
Michael Moon	(The Mount School, York)	IIIMC / JMO
Peter Neumann	(Oxford University)	IIIMC
Sylvia Neumann		IIIMC
Jenny Ramsden		JMO
Chris Robson	(University of Leeds)	IIIMC
Lyn Robson		IIIMC
Bill Richardson	(Elgin Academy)	JMO
Alan Slomson	(University of Leeds)	IIIMC / JMO
Patricia Smart	(Dean Close School, Cheltenham)	IIIMC
Alex Voice	(St Christopher's School, Hove)	JMO
John Webb	(University of Cape Town)	JMO

Problems Groups

There are four groups. The BMO group is:

Christopher Bradley (Clifton College)
Kevin Buzzard (Imperial Coll. London)
Tim Cross (KES, Birmingham)
Vin de Silva (Trinity Coll. Cambridge)
Ben Green (Trinity College, Cambridge)
Gerry Leversha (St Paul's School)
David Monk (ex Edinburgh University)
Peter Shiu (Loughborough University)

The other three groups have overlapping membership. There is one group for each of the Senior Mathematical Challenge, the Junior and Intermediate Mathematical Challenges and the Junior Mathematical Olympiad. As the chair, Howard Groves was involved in all of these three. Others involved are as shown. (S = Senior Challenge; I&J = Intermediate and Junior Challenges; JMO = Junior Mathematical Olympiad.)

Tony Gardiner	(University of Birmingham)	JMO
Colin Dixon	(ex RGS Newcastle)	
Mark Ford	(Yarm School)	
Terry Heard	(ex City of London School)	I & J / JMO
Margaret Jackson	(ex Tormead School)	I & J
Susie Jameson	(Wells Cathedral School)	I & J
Andrew Jobbings	(Bradford Grammar School)	S / I & J
Peter Ransom	(The Mountbatten School)	I & J
Bill Richardson	(Elgin Academy)	JMO
Andrew Rogers	(King Edward VI, Camphill)	JMO
Steve Rout	(Queen Mary GS, Walsall)	JMO
Alan Slomson	(University of Leeds)	S / I & J
Patricia Smart	(Dean Close School, Cheltenham)	S
Alex Voice	(St Christopher's School, Hove)	I & J

Residential Schools

Trinity Training Session

Richard Atkins (Oundle School)
Kevin Buzzard (Imperial Coll. London)
Dan Crisan (Queens' Coll. Cambridge)
Tony Gardiner (Univ. of Birmingham)
Julia Gog (Trinity College, Camb.)
Ben Green (Trinity College, Camb.)
Imre Leader (University Coll. London)
Adam McBride (Univ. of Strathclyde)
David Monk (ex Univ. of Edinburgh)
Christopher Bradley (Clifton College)

Birmingham Summer School

Richard Atkins (Oundle School)
Christopher Bradley (Clifton Coll.)
Julian Gilbey (QMW, London)
Ben Green (Trinity College, Camb.)
Howard Groves (RGS, Worcester)
Nick Lord (Tonbridge School)
Adam McBride (Univ. of Strathclyde)
Dan Crisan (Queen's Coll. Cambridge)
Imre Leader (University Coll. London)
Chris Robson (University of Leeds)
Angela Gould (UKMT)
Mary Teresa Fyfe (Hutchesons' GS)

UKMT Yearbook Order Form

To order more copies of this Yearbook, please complete the following and send with a cheque for £5 per copy payable to "UKMT Leeds" to:

Ms Jenny Gill
Maths Challenges Office
School of Mathematics
University of Leeds
Leeds LS2 9JT

I would like to order ..... copies of the Yearbook at £5 per copy. I have enclosed a cheque for £ .....

Title .... Initial ....

Surname ....

Address ....

....

Postcode ....

Telephone ....

United Kingdom Mathematics Trust – a registered charity